Electrophysiology for Clinicians

EDITED BY

MIGUEL A. BARRERO GARCIA MD
Clinical Lecturer, Université de Montréal
Cardiologist, Centre Hospitalier Régional de Trois Rivières, Quebec, Canada

PAUL KHAIRY MD, PhD, FRCP(C)
Canada Research Chair, Electrophysiology and Adult Congenital Heart Disease
Director, Adult Congenital Heart Center, Montreal Heart Institute
Associate Professor of Medicine, Université de Montréal, Quebec, Canada
Boston Adult Congenital Heart Service, Harvard University, Boston

LAURENT MACLE MD, FRCP(C)
Associate Professor of Medicine, Université de Montréal
Cardiologist and Director of Electrophysiology Fellowship Training Program,
Montreal Heart Institute, Quebec, Canada

STANLEY NATTEL MD
Professor of Medicine and Paul-David Chair in Cardiovascular Electrophysiology,
Université de Montréal
Cardiologist and Director of Electrophysiology Research Program,
Montreal Heart Institute, Quebec, Canada

Minneapolis, Minnesota

Cardiotext Publishing, LLC
3405 W. 44th Street
Minneapolis, Minnesota 55410
USA

www.cardiotextpublishing.com

Any updates to this book may be found at:
www.cardiotextpublishing.com/titles/detail/9781935395140

Comments, inquiries, and requests for bulk sales can be directed to the publisher at:
info@cardiotextpublishing.com.

ISBN-13: 978-1-935395-14-0

Library of Congress Control Number: 2011924286

Printed in the United States of America

Contents

PART 3: SYNCOPE, SUDDEN CARDIAC DEATH, AND CLINICAL ARRHYTHMIAS

PART 4: PACEMAKERS, DEFIBRILLATORS, AND TECHNIQUES

About the Contributors

Editors

Miguel A. Barrero Garcia MD
Clinical Lecturer, Université de Montréal
Cardiologist, Centre Hospitalier Régional de Trois Rivières, Quebec, Canada

Paul Khairy MD, PhD, FRCP(C)
Canada Research Chair, Electrophysiology and Adult Congenital Heart Disease
Director, Adult Congenital Heart Center, Montreal Heart Institute
Associate Professor of Medicine, Université de Montréal, Quebec, Canada
Boston Adult Congenital Heart Service, Harvard University, Boston

Laurent Macle MD, FRCP(C)
Associate Professor of Medicine, Université de Montréal
Cardiologist and Director of Electrophysiology Fellowship Training Program, Montreal Heart Institute, Quebec, Canada

Stanley Nattel MD
Professor of Medicine and Paul-David Chair in Cardiovascular Electrophysiology, Université de Montréal
Cardiologist and Director of Electrophysiology Research Program, Montreal Heart Institute, Quebec, Canada

Contributors

Marc Dubuc MD, FRCP(C), FACC, FHRS
Electrophysiology Service, Department of Medicine, Montreal Heart Institute
Associate Professor of Medicine, Université de Montréal, Montreal, Quebec, Canada

Peter G. Guerra MD
Chief of Cardiac Electrophysiology, Electrophysiology Service, Department of Medicine, Montreal Heart Institute
Assistant Professor of Medicine, Université de Montréal, Montreal, Quebec, Canada

Léna Rivard MD, FRCP(C)
Electrophysiology Service, Department of Medicine, Montreal Heart Institute, Université de Montréal, Montreal, Quebec, Canada

Mario Talajic MD, FRCP(C), FACC, FHRS
Electrophysiology Service, Department of Medicine, Montreal Heart Institute
Chair, Department of Medicine, Université de Montréal, Montreal, Quebec, Canada

Bernard Thibault MD, FACC, FHRS
Electrophysiology Service, Department of Cardiology, Montreal Heart Institute
Assistant Professor of Medicine, Université de Montréal, Montreal, Quebec, Canada

Foreword

William G. Stevenson, MD
Professor of Medicine
Harvard Medical School
Director, Cardiac Arrhythmia Program
Brigham and Women's Hospital
Boston, Massachusetts

Cardiac arrhythmias present major management issues for all physicians who care for patients who have heart disease. The development and now common use of invasive electrophysiologic techniques, often in concert with catheter ablation, has fostered an explosion of knowledge in this field. The resulting definitions of mechanisms and physiology have translated into nuanced understandings of clinical features of common cardiac arrhythmias, as well as rare ones. Detailed cardiac mapping studies have defined new varieties and types of arrhythmias, and the findings from these studies have been related back to the electrocardiogram and other clinical features of arrhythmias. For example, a variety of types of atrial flutter have been defined, and specific types can be anticipated from the nature of the underlying heart disease, and from the electrocardiogram. This recognition is not merely of academic interest, but also informs the clinician about the likely success, ease, and risks of ablation therapy for that arrhythmia, information important to assessing treatment options. The situation is similar for other supraventricular tachycardias, as well as for ventricular tachycardias.

A number of genetic syndromes causing ventricular arrhythmias and sudden death are now recognized. The common use of implantable defibrillators has improved our understanding of the nature of ventricular arrhythmias that cause sudden death in rare syndromes and common cardiac diseases. All physicians will encounter patients for whom these devices can be life-saving. The optimal use of implantable devices involves not only recognition of the patients who need them, but also awareness of the significant progress in strategies to avoid potential adverse effects of these devices, such as the potential for diminished cardiac function during right ventricular pacing, and inappropriate shocks that reduce quality of life.

While this expanding body of knowledge is disseminated among subspecialists at the cutting edge of the field, it is often not organized and presented in an easily accessible format for the vast majority of health care providers. *Electrophysiology for Clinicians* is a superb distillation of the field for clinicians. Authored by leaders in the field, led from the Montreal Heart Institute, it is a clear and concise text emphasizing clinically valuable insights and providing their pathophysiologic basis. Overviews of the fundamentals of arrhythmias and therapies provide the clinician with the necessary foundation for incorporating and retaining new advances into his or her knowledge base. This book is of great value to health care providers who care for patients who have cardiac arrhythmias.

Preface

The study of rhythm and conduction abnormalities in modern times has been facilitated by sophisticated diagnostic and therapeutic techniques. The explosion of knowledge regarding mechanisms that govern rhythm abnormalities has led to the creation of a highly specialized field, and consequent evolution of new concepts, terminology, and approaches.

Despite technological advances bringing increasingly complex interventions, the backbone of diagnostic and management strategies in cardiac electrophysiology rests on clinical grounds. Unfortunately, textbooks, monographs, and reviews largely focus on technical aspects or rely on ultraspecific electrophysiological jargon, limiting their utility as teaching tools for health professionals interested in clinical applications of state-of-the-art knowledge. This book is a response to the requests of clinicians, trainees, and other health professionals for essential information in a concise, readable, and easily accessible format.

Electrophysiology for Clinicians attempts to unravel the complexities inherent to cardiac electrophysiology, with a particular focus on diagnosis and management of cardiac arrhythmias, and indications for patient referral. With its standardized chapter outline approach including key points, color figures, charts, and tables, this handy guide will help demystify the essentials of cardiac electrophysiology for the practicing caregiver.

Acknowledgments

To my wife, Annia, and to our children Miguel, Manuel, and Marcel.

I also wish to acknowledge Sebastien Grefford for his personal assistance in making this book possible.

Miguel A. Barrero Garcia

Abbreviations

AAD	antiarrhythmic drug
ACEI	angiotensin-converting enzyme inhibitors
ACLS	advanced cardiac life support
AH	atrium-His
AF	atrial fibrillation
AFL	atrial flutter
AP	action potential
APD	AP duration
ARVC/D	arrhythmogenic RV cardiomyopathy/dysplasia
AT	atrial tachycardia
ATP	antitachycardia pacing
AV	atrioventricular
AVN	AV node
AVNRT	AV nodal reentrant tachycardia
AVRT	AV reentrant tachycardia
BBB	bundle branch block
BBR	bundle branch reentry
BLS	basic life support
BP	blood pressure
BrS	Brugada syndrome
BRS	baroreflex sensitivity
CAD	coronary artery disease
CCB	Ca^{2+} channel blockers
CHD	congenital heart disease
CICR	Ca^{2+}-induced Ca^{2+}-release
CL	cycle length
CPVT	cathecolaminergic polymorphic VT
CrCl	creatinine clearance
CRT	cardiac resynchronization therapy
CSM	carotid sinus massage
DCM	dilated cardiomyopathy
DFT	defibrillation threshold

ECG	electrocardiogram
ECV	electric cardioversion
EF	ejection fraction
EGM	electrogram
EP	electrophysiological
EPS	EP study
EST	exercise stress testing
FAT	focal AT
HCM	hypertrophic cardiomyopathy
HF	heart failure
HR	heart rate
HRV	heart rate variability
HRT	heart rate turbulence
HV	His-ventricle
ICD	implantable cardioverter defibrillator
IHR	intrinsic heart rate
INR	international normalized ratio
IST	inappropriate sinus tachycardia
IV	intravenous
LA	left atrium
LBBB	left BBB
LMWH	low molecular-weight heparin
LQTS	long QT syndrome
LV	left ventricle
LVH	LV hypertrophy
MAT	multifocal AT
MI	myocardial infarction
NIDCM	nonischemic DCM
NSR	normal sinus rhythm
NSTEMI	non ST-elevation myocardial infarction
NSVT	nonsustained VT
PAC	premature atrial contraction
PJRT	permanent junctional reciprocating tachycardia
PMT	pacemaker-mediated tachycardia
POTS	postural orthostatic tachycardia syndrome
PVAB	postventricular atrial blanking
PVARP	postventricular atrial RP
PVC	premature ventricular contraction
PWM	P-wave morphology
RA	right atrium
RBBB	right BBB

RP	refractory period; RP interval
SA	sinoatrial
SAECG	signal-averaged ECG
SAN	sinoatrial node
SCD	sudden cardiac death
SHD	structural heart disease
SND	sinus node dysfunction
SNRT	sinus node reentrant tachycardia
SMVT	sustained monomorphic VT
SR	sarcoplasmic reticulum
SSS	sick sinus syndrome
STEMI	ST-elevation myocardial infarction
SVT	supraventricular tachycardia
TARP	total atrial RP
TdP	torsade de pointes
TTT	tilt table test
TWA	T-wave alternans
ULV	upper limit of vulnerability
VA	ventriculoatrial
VF	ventricular fibrillation
VT	ventricular tachycardia
WPW	Wolff-Parkinson-White

PART 1

Basic Concepts

Chapter 1

Basic Cardiac Electrophysiology

Miguel A. Barrero Garcia and Stanley Nattel

I. THE ELECTRICAL SYSTEM OF THE HEART

Generation of the heart rhythm takes place in the sinoatrial node (SAN), the normal pacemaker of the heart: from this region, spontaneous action potentials (APs) propagate to the rest of the cardiac tissue, possibly through specialized conducting pathways (Figure 1.1). The ordered propagation and activation of the heart permits the synchronous contraction of the different cardiac chambers and prevents excessively slow or excessively rapid heart rate (HR).

Cells of the SAN have the greatest automaticity, which means they possess the highest intrinsic rate (i.e., the highest rate of spontaneous discharge of a region of the heart when it is acting as the cardiac pacemaker). SAN cells exert overdrive suppression of subsidiary potential pacemaker tissue (e.g., cells in the atrioventricular node [AVN] and His-Purkinje cells). Cells of the SAN do not have a true resting membrane potential: their transmembrane voltage constantly cycles from around –50 mV (more positive than that of other cells) to about +10 mV. They are depolarized relative to other cardiac tissue because of an absence of background inward rectifier K^+ I_{K1} channels. In contrast, myocytes in atrial and ventricular tissues have a true resting membrane potential, usually at around –80 mV (this negative value is the result of the dominant effect of a steady net efflux of K^+ by way of I_{K1}).

Some other significant differences within this specialized conducting system are:

- The AP upstroke is faster in atrial and ventricular cells, owing to the fact that a large and fast Na^+ current dominates during phase 0. A smaller L-type Ca^{2+} current ($I_{Ca(L)}$) dominates the upstroke in pacemaker (SAN) cells.
- A longer phase 2 in the ventricular AP makes the ventricular tissue refractory to overly rapid reexcitation, avoiding tetany.

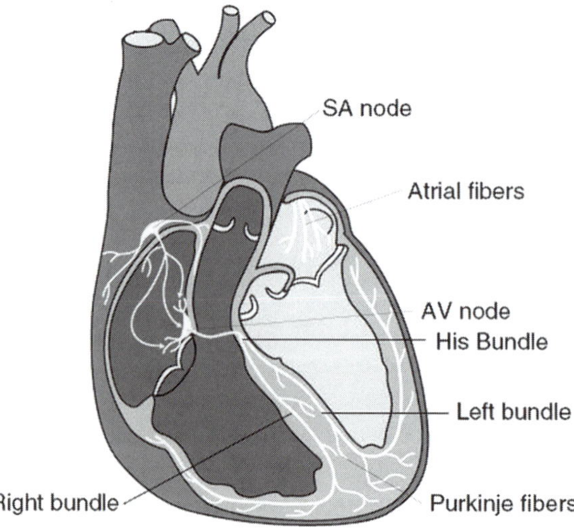

Figure 1.1. The specialized cardiac conducting system. © Medtronic, Inc. 2010.

- Cells of the AVN lack rapid Na^+ channels; their depolarization is slower, mainly depending on slow Ca^{2+} channels (see following).
- Sympathetic and parasympathetic fibers richly innervate SAN, atrial, and AVN cells, whereas in the ventricles the parasympathetic innervation is sparse. Parasympathetic influences hyperpolarize and abbreviate APs, decreasing the refractory period (RP) and suppressing automaticity by moving maximum negative potential farther away from threshold. Sympathetic innervation increases automaticity and RP, while exerting potentially complex effects on AP duration because of the multiple currents affected.

For the AP to be propagated, excited cardiac cells provide the electrical current needed to depolarize neighboring unexcited cells to excitation threshold. This transmission of current occurs via intercellular low-resistive pathways known as gap junctions.

The success of conduction also depends on the excitability of the cell membrane, which in turn depends on the maximum excitatory current and the voltage difference between resting potential and the threshold for firing. When the source current of excited cells is small relative to the current sink of cells needing to be excited (e.g., at branch points or regions of poor coupling), very slow conduction can occur. Because of the source-sink relationship, curvature of the electrical wavefront (e.g., in the presence of functional zones of block or structural obstacles) affects the ability to propagate, with the propagation velocity depending on the radius of the curvature (conduction is not sustained below a critically small radius). As can be inferred, these interactions are of particular importance for explaining the mechanisms of conduction slowing and/or block and reentrant arrhythmias.

At a macroscopic level, propagation of the AP follows an anisotropic pattern, influenced by cell size/shape (larger, elongated, anisotropically aligned cells conduct faster), location of gap junctions (much more concentrated at cell ends, in the longitudinal direction, than at lateral margins governing transverse propagation), and ion channel clustering around gap junctions (which could facilitate conduction when gap junction coupling is greatly reduced). As a result of all these characteristics there is great variation in AP behavior within a particular cardiac region, among specific cardiac regions, and obviously within the cardiac conduction system itself.

Once cardiac tissue has been activated, electromechanical coupling initiates mechanical activity. Contraction results when Ca^{2+} entry triggers a secondary release of Ca^{2+} from the sarcoplasmic reticulum (Ca^{2+}-induced Ca^{2+}-release, or CICR); this increased cytosolic Ca^{2+} removes inhibition of actin-myosin interaction, producing contraction.

Contraction is terminated by the active removal of Ca^{2+} from cytosol back into the sarcoplasmic reticulum (SR) by an active pump (the SR Ca^{2+}-ATPase, or SERCA). Once cardiac tissue is activated, it remains unavailable for reactivation during a finite period of time (the RP).

II. ION MECHANISMS OF THE CARDIAC ACTION POTENTIAL

The AP is controlled by the movement of selected ions through cell membrane voltage-gated ion channels, pumps, and exchangers; this movement creates electrical potentials in and out of the cells, with an equilibrium potential at which there is no net force for a given current to flow across the membrane. The direction of flow, and therefore of electrical current generation, is determined by the transmembrane concentration gradient (ion concentration outside versus inside the cell, Table 1.1), together with the electrical gradient resulting from the transmembrane potential.

When the membrane potential in cardiac cells is stimulated above its threshold value, a positive shift in the membrane potential (*depolarization*) occurs, caused by opening of ion channels that allows positively charged Na^+ and Ca^{2+} to enter the cell; after its peak, the membrane potential is restored to its original value during the *repolarization* phase, largely due to opening of K^+ channels. This cycle of electrical activity, the AP, is commonly described in phases (Figure 1.2):

- **Phase 4:** Nonautomatic cells stay at the stable resting membrane potential of the cell, dominated by the inwardly rectifying K^+ current (I_{K1}). Pacemaker cells do not have a stable resting potential; following phase 3 repolarization, they gradually depolarize in phase 4 as illustrated by the dashed lines in Figure 1.2.
- **Phase 0:** Rapid depolarization phase, due to a sudden inward increase in ion permeability, mainly due to opening of fast Na^+ channels; this potential peaks at around +40 mV.
- **Phase 1:** Occurs due to inactivation of the fast Na^+ channels, followed by activation of I_{to}, which results in a transient net outward current of K^+ and inward movement of Cl^- (carried by I_{to1} and I_{to2}, respectively), leading to rapid repolarization.

Table 1.1. Intra- and Extracellular Ion Concentrations (mmol/L)

Ion	Extracellular	Intracellular	Ratio
Na^+	135–145	10	14:1
K^+	3.5–5	155	1:16
Cl^-	95–110	20–30	4:1
Ca^{2+}	2	10^{-4}*	2×10^{-4}

*Free cytosolic concentration varies from the range of 1×10^{-7} to 4×10^{-7} mmol/L over the course of a cardiac cycle.

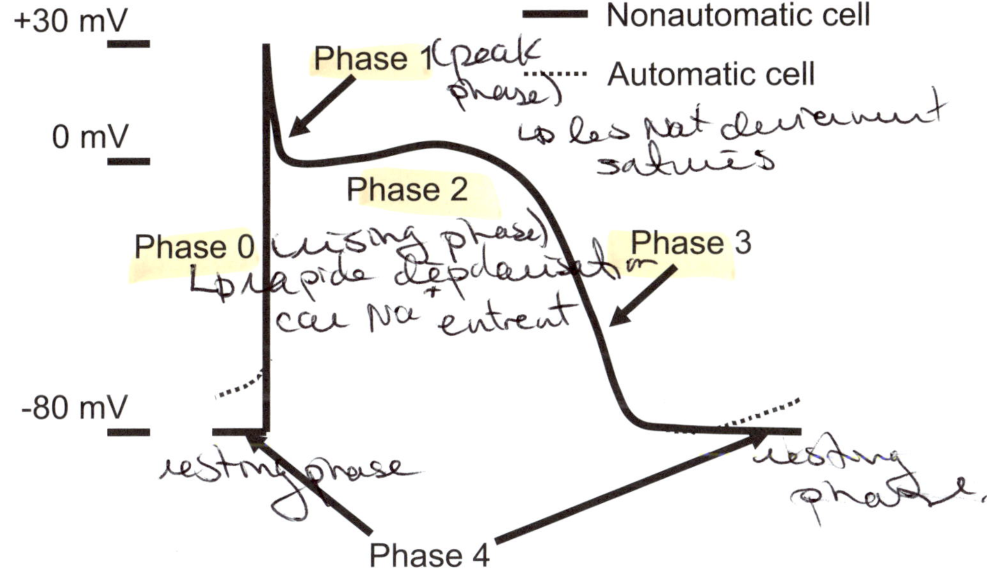

Figure 1.2. Phases of the ventricular AP. Courtesy of Dr. Stanley Nattel.

- **Phase 2:** Also known as the plateau phase, at around +20 mV: no particular conductance is dominant as there is a balance between inward movement of Ca^{2+} through L-type Ca^{2+} channels ($I_{Ca(L)}$) and outward movement of K^+ through rapid ($I_{K,r}$) and slow (I_{Ks}) delayed rectifier K^+ channels. The Na^+/Ca^{2+} exchanger current ($I_{Na, Ca}$) and the Na^+/K^+ pump current ($I_{Na, K}$) also play minor roles during this phase.
- **Phase 3:** The prolonged duration of the plateau phase allows for eventual activation of I_{Ks} and $I_{K,r}$ (while L-type Ca^{2+} channels close), provoking a net outward current that changes negatively the membrane potential, allowing further K^+ channel opening (primarily I_{Kr}, and the inwardly rectifying K^+ current, I_{K1}), with subsequent repolarization and return to resting membrane potential. I_{K1} remains open throughout phase 4, setting the resting membrane potential.

The I_f current seems to play a major role in the spontaneous initiation of diastolic depolarization and in the chronotropic regulation of the HR, modulated by autonomic neurotransmitters. This current is carried by Na^+ or K^+, and its equilibrium potential is intermediate (~ –25 mV). I_f activates slowly upon repolarization, and then starts to depolarize the cell toward its equilibrium potential. The cell types with the most I_f are SAN, AVN, and Purkinje tissue, in that order. Working atrial and ventricular muscle have much less I_f. Pacemaking activity has dual mechanisms, including I_f-dependent depolarization ("membrane clock") and a complex Ca^{2+}-cycling system between $I_{Ca(L)}$, the SR, and various Ca^{2+} pumps and exchangers ("Ca^{2+}-clock"). This dual system acts as a safeguard for the crucially important pace-making function of the heart.

III. REGULATION OF IONIC PERMEABILITIES

Ion channels, apart from being selective—that is, different ions have their specific channels (Table 1.2)—generally need to be activated to allow a current to pass.

Channels are generally closed at rest (except for the non-voltage-gated current I_{K1}), but when an electrical stimulus is applied they undergo a conformational change that allows current to flow, in a process called *activation*. Some voltage-dependent channels also undergo further changes if the stimulus is maintained so that, while the channel remains fully opened, it becomes poorly conducting; this process (voltage- and time-dependent, as with activation) is termed *inactivation*. As a result of this "voltage gating," the function of Na^+ channels exists in 3 states: open channels (selectively allow Na^+ to traverse the membrane), resting channels (impermeable to Na^+ but are available for conversion to the open state), and inactivated channels (impermeable to Na^+, they have to enter a resting state before they can open).

Since Na^+ channels have to get back to –60 mV to recover, after which recovery is very rapid, the AP duration (APD) is the main determinant of the RP in

Table 1.2. Major Currents during the Ventricular AP

Ion	Current	α Subunit Protein	α Subunit Gene	Phase/Role
Na^+	I_{Na}	$Na_V1.5$	SCN5A	0
Ca^{2+}	$I_{Ca(L)}$	$Ca_V1.2$	CACNA1C	0–2
K^+	I_{to1}	$K_V4.2/4.3$	KCND2/KCND3	1, notch
K^+	I_{Ks}	$K_V7.1$	KCNQ1	2, 3
K^+	I_{Kr}	$K_V11.1$ (hERG)	KCNH2	3
K^+	I_{Kl}	$K_{ir}2.1/2.2/2.3$	KCNJ2/KCNJ12/ KCNJ4	3, 4
Na^+, Ca^{2+}	$I_{Na, Ca}$	$3Na^+$-$1Ca^{2+}$-exchanger	NCX1 (SLC8A1)	Ion homeostasis
Na^+, K^+	$I_{Na, K}$	$3Na^+$-$2K^+$-ATPase	ATP1A	Ion homeostasis
Ca^{2+}	$I_{p, Ca}$	Ca^{2+}-transporting ATPase	ATP1B	Ion homeostasis

fast-channel tissue (atria, ventricles, His-Purkinje system). Because APD decreases with increased HR, this leads to the RP decreasing in fast-channel tissue as HR increases. Slow-channel tissues (sinus and AVN cells) differ from fast-channel tissues because the depolarized resting potentials in these regions (positive to –60 mV) inactivate Na^+ channels. These areas are therefore totally dependent on a second channel, the "slow inward" $I_{Ca(L)}$ channel (carrying Ca^{2+} as the inward ion for excitation and conduction) for phase 0 depolarization. The main determinant of their RP is the slow recovery from inactivation of Ca^{2+} channels and not the duration of the AP per se. Thus, the RP in the AVN does not decrease significantly when the AVN is activated rapidly during atrial tachycardias. This phenomenon helps to limit the number of atrial impulses that are conducted through the AVN during rapid atrial arrhythmias (e.g., atrial fibrillation [AF]), contributing importantly to the protective "filter" function of the AVN.

Other mechanisms involved in transmembrane ionic movement are active pumping, which generates continuous low-level movement of ions (e.g., Na^+, K^+-ATP'ase pumps Na^+ out of the cell and K^+ into the cell, with the movement of these ions coupled); and exchange mechanisms (e.g., Na^+, Ca^{2+} exchange, which can exchange Na^+ for Ca^{2+} across the membrane in either direction).

Genetic mutations may lead to alterations of the currents and phases/roles ion channels codify; these changes form the basis for several known ion channelopathies (or inherited arrhythmia syndromes).

The existence of distinct ion channels, with different roles, is the basis for the design and classification of antiarrhythmic drugs. These medications bind to closed (resting), open (activated), or inactivated channels, but their affinity for resting channels is usually very low. Drugs that bind to open or inactivated channels show *positive use dependency*, a greater inhibitory action at faster HR. *Reverse use dependency* refers to the fact that many APD-prolonging drugs show their greatest effects at slow HR, and is not directly related to the differential affinity of drugs for different channel states.

SUGGESTED READINGS

Barbuti A, Baruscotti M, Difrancesco D. The pacemaker current: from basics to the clinics. *J Cardiovasc Electrophysiol.* 2007;18:342-347.

Dobrzynski H, Boyett MR, Anderson RH. New insights into pacemaker activity: promoting understanding of sick sinus syndrome. *Circulation.* 2007;115:1921-1932.

Kleber AG, Rudy Y. Basic mechanisms of cardiac impulse propagation and associated arrhythmias. *Physiol Rev.* 2004;84:431-488.

Nattel S, Maguy A, Le Bouter S, Yeh Y-H. Arrhythmogenic ion-channel remodeling in the heart: heart failure, myocardial infarction, and atrial fibrillation. *Physiol Rev.* 2007;87:425-456.

Chapter 2

Antiarrhythmic Drugs

Miguel A. Barrero Garcia and Stanley Nattel

I. GENERAL PRINCIPLES

Antiarrhythmic drugs (AADs) are medications designed to terminate cardiac arrhythmias and prevent their recurrence. Most classical AADs work by blocking ion channels in cardiac myocytes or by interacting with membrane receptors that modulate cardiac bioelectricity.

The most widely used AAD classification system is the modified Vaughn Williams classification (Table 2.1), mainly because of its simplicity and ability to predict AAD efficacy/adverse reaction profile. This classification categorizes AADs based on principal effects on the action potential (AP); however, many drugs have mixed actions and some active metabolites may even have significant effects different from the parent compound. The Sicilian Gambit, a more comprehensive AAD classification system, has not found its way into clinical practice due to the complexity of its use.

The effects of AADs should be monitored closely, both for the purpose of determining efficacy and to assess the risk of proarrhythmia: this monitoring should be done not only based on symptoms but also directly with serial electrocardiograms (ECGs), Holter monitoring, event loop recorders, and/or exercise stress testing (EST).

A. Modified Vaughn Williams Classification

1. Class I drugs

Drugs belonging to class I of the modified Vaughn Williams classification block fast Na^+ channels, decreasing the rate of phase 0 depolarization of the AP and slowing the rate of impulse conduction throughout fast-channel tissue (atrial and ventricular muscle, His-Purkinje system). Many of these drugs also block K^+ channels, with varying effect on AP duration (APD). As a result of Na^+ channel blockade, they possess local anesthetic and membrane-stabilizing effects. They are further classified into:

a. Class Ia (quinidine, procainamide, disopyramide): Class Ia AADs are moderately potent Na^+ channel blockers that slow the AP upstroke, increasing APD thanks to a potent K^+-channel-blocking effect; they also decrease conduction speed and prolong refractoriness of fast-channel tissue. Their major indication is the treatment of atrial tachyarrhythmias. There is a moderate risk of proarrhythmia, due to facilitated reentry secondary to slowed conduction or torsade de pointes (TdP) due to prolonged APD/QT interval. Relatively safe in patients in patients with implantable cardioverter defibrillators (ICDs), they are contraindicated if structural heart disease (SHD) (including left ventricle hypertrophy [LVH]) or active coronary artery disease (CAD) is present.

Table 2.1. Modified Vaughan Williams Classification

Class	Basic Mechanism	EP Effect
I	Na^+ channel blockade	Reduce phase 0 slope and peak of AP
Ia	Moderate effect	Moderate reduction phase 0 slope; ↑APD and effective RP
Ib	Weak effect	Small reduction phase 0 slope; ↓APD and effective RP
Ic	Strong effect	Pronounced reduction phase 0 slope; no effect on APD or effective RP
II	β-blockade	Block sympathetic activity; ↓rate and AVN conduction
III	K^+ channel blockade	Delay in repolarization (phase 3); ↑APD and effective RP
IV	Ca^{2+} channel blockade	Block of L-type Ca^{2+} channels; more effective in SAN and AVN; ↓rate and conduction
V	Varied	↓rate and conduction

b. Class Ib (lidocaine, mexiletine, tocainide, phenytoin): Class Ib agents are fast-unblocking Na^+ channel blockers, with primarily voltage-dependent Na^+ channel block, little effect on phase 0 upstroke, but substantial Na^+ channel block during the depolarized phases of the AP and during spontaneous phase 4 depolarization. They increase the threshold for excitability, decrease automaticity in fast-channel tissue by decreasing the slope of phase 4, and decrease APD. Their effect is greatest in depolarized tissue or at rapid heart rate (HR). There is a low risk of proarrhythmia (usually reentry), and QT prolongation almost never occurs.

c. Class Ic (flecainide, propafenone): Class Ic agents are very potent Na^+ channel blockers, with some K^+-channel-blocking activity. At controlled rates, class Ic AADs have no net effect in APD, as their Na^+-blocking effects offset any APD-prolonging effects of their weak K^+-channel-blocking actions. They prolong the QRS complex at a normal HR by blocking Na^+ channels and slowing conduction, with some lengthening of the QT interval due to QRS prolongation. A significant increase in APD in the atria at faster rates makes them useful in atrial tachyarrhythmias and atrial fibrillation (AF). Propafenone has modest β-blocking effects. There is an increased risk of proarrhythmia, usually due to reentry. Safe in the absence of SHD, Ic drugs are contraindicated in patients with history of myocardial infarction (MI), sustained ventricular arrhythmias, or left ventricle (LV) dysfunction. Because they can increase the ventricular rate in patients with atrial flutter (AFL), it is always advisable to combine them with AVN-blocking agents.

2. Class II drugs (β-blockers)

The antiarrhythmic effect of class II agents is indirect, by postsynaptic β-adrenergic blockade mainly on the sinoatrial (SA) and atrioventricular (AV) nodes (SAN and AVN), preventing the activation of the adenylate cyclase and the increase in intracellular cAMP that occurs in response to enhanced sympathetic activation, which results in decreased automaticity and slowing of AVN conduction. Despite this lack of direct electrophysiological (EP) effect they have proved to be effective in preventing sudden cardiac death (SCD), probably by improving the myocardial oxygen supply-demand relationship and blunting the response to catecholamines. They are also of utility after MI, in controlling the ventricular rate in supraventricular tachycardia (SVT), preventing torsade de pointes (TdP) in congenital long QT syndrome (LQTS), and decreasing the recurrence of neurocardiogenic syncope.

3. Class III drugs (amiodarone, bretylium, sotalol, ibutilide, dofetilide, azimilide, dronedarone)

Class III agents have a direct effect to increase APD (greater at slow HR, so-called reverse use-dependent effect) and refractoriness by blocking K^+ channels. There is some risk of TdP due to early afterdepolarizations and associated reentry related to excess spatially heterogeneous APD prolongation.

4. Class IV drugs (Ca^{2+} channel blockers)

The Ca^{2+} channel blockers (CCBs) block L-type Ca^{2+} channels, with effects manifest mainly in the SANs and AVNs, where they induce concentration-dependent depression in phase 4 depolarization and a prolongation in refractoriness, which results in depressed automaticity and slowed conduction.

5. Class V drugs

The original Vaughan Williams classification does not include drugs like digoxin and adenosine, with antiarrhythmic actions exerted on arrhythmias involving the AVN. Digoxin suppresses AV nodal conduction by enhancing vagal effects on the heart, whereas adenosine acts by enhancing the adenosine-sensitive K^+ current, I_{KAdo}. I_{KAdo} is carried by the same ion channels as I_{KACh}, which mediates vagal effects on the heart, so that the ionic mechanism of digitalis and adenosine effects is similar. The main difference between the 2 types of compounds is in the speed of onset and offset of action. Adenosine effects are very short-lived, so it is given intravenously to terminate AVN reentrant tachycardias. Digoxin effects are slow to develop and dissipate, so the drug is mainly used chronically to control the ventricular response to AF.

II. SPECIFIC ANTIARRHYTHMIC DRUGS

A. Class Ia Drugs

1. Quinidine

Quinidine exerts a direct action by blocking Na^+ channels, depressing automaticity of ectopic foci (therefore suppressing premature atrial contractions [PACs] and premature ventricular contractions [PVCs]). It also slows conduction and increases refractoriness, converts AF to sinus rhythm and prevents its recurrence, and terminates and prevents SVTs. Additionally it has an indirect action via its anticholinergic effects, interfering with vagal innervation to the heart. In a minority of patients it may be used to prevent recurrence of ventricular tachycardia/ventricular fibrillarion (VT/VF) (e.g., Brugada syndrome [BrS] patients). It does not reduce the incidence of SCD or improve survival.

a. Dose: 200 to 300 mg PO q8h (maximum 600 mg q8h). Extended-release tablets (quinidine gluconate) are given q8–12h.

b. Pharmacokinetics and metabolism: Oral absorption is almost complete, and it is metabolized primarily in the liver. The serum half-life is 5 to 8 hours, increasing significantly with age and hepatic dysfunction. Serum levels of 1.3 to 5.0 mg/L correlate with clinical efficacy.

c. Special precautions/toxicities: Due to quinidine's vagolytic and atrial conduction-slowing effects (which slow the atrial rate in AF), control of ventricular rate in AF/AFL should be achieved with AVN-blocking agents before prescribing it. Gastrointestinal side effects are common and frequently force its suspension. Proarrhythmia is seen in 1% to 3% of cases (quinidine should be avoided if there is a history of TdP and discontinued if QTc exceeds 500 milliseconds or increases > 25% from baseline values). New or increased AV conduction abnormalities may occur. Other possible adverse reactions include central nervous system toxicity, anemia, thrombocytopenia, hepatotoxicity, hypotension, and autoimmune disorders. Drug interactions can also occur: for example, warfarin and digoxin effects are potentiated by quinidine, while phenobarbital and phenytoin reduce its half-life significantly.

2. Procainamide

Procainamide has similar pharmacological effects and clinical utility to quinidine, but it does not prolong the QRS in therapeutic concentrations. Its major metabolite, N-acetylprocainamide (NAPA), has class III properties. It has long been considered the drug of choice in the acute treatment of AF in Wolff-Parkinson-White (WPW) syndrome (currently amiodarone is often preferred). Procainamide is better than lidocaine for the treatment of sustained monomorphic ventricular tachycardia (SMVT).

a. Dose: Oral loading dose is usually 600 mg to 1g (about 12 mg/kg), with maintenance of 2 to 4 g/day (alternate dosing is 50 mg/kg/day, not exceeding 4 g/day). Intravenous (IV) procainamide is given with a loading dose of 750 mg to 1g (given no faster than 20 mg/min), continuing until arrhythmia is suppressed, hypotension ensues, QRS widens by 50%, or a maximum loading dose of 17 mg/kg is administered; maintenance is usually at 1 to 4 mg/min.

b. Pharmacokinetics and metabolism: Peak concentration is almost immediate after IV administration, but takes about 1 hour after an oral dose. The drug is biotransformed in the liver by a genetically controlled acetylation system: slow acetylators have longer procainamide half-lives and produce less NAPA. Serum half-life is about 3 hours (6–8 hours for NAPA), and increases if hepatic or renal dysfunction is present. Serum levels of 4 to 12 mg/L correlate with clinical efficacy.

c. Special precautions/toxicities: A lupuslike syndrome occurs in up to one-third of patients with chronic use (mainly in hepatic slow acetylators) and resolves when the medication is stopped. Antinuclear antibodies appear in 75% of cases, not indicating the need to stop the drug if no other symptoms are present. Proarrhythmia and ECG changes are similar to that of quinidine, and continuous ECG and blood pressure (BP) monitoring are suggested during IV administration.

3. Dysopiramide

Disopyramide shares similar EP effects with other class Ia agents. It is most often used in the treatment of ventricular arrhythmias and may be beneficial for obstructive hypertrophic cardiomyopathy (HCM).

a. Dose: 100 mg PO q6h, up to 300 mg q6h. In cases of hepatic impairment, the dose should not exceed 100 mg q6h, whereas if renal impairment exists the dose should be titrated according to creatinine clearance (CrCl 40–80 mL/min: 100 mg q6h; 30–40 mL/min: 100 mg q8h; 15–30 mL/min: 100 mg q12h; < 15 mL/min: 100 mg q24h).

b. Pharmacokinetics: Peak plasma concentration is reached in about 2 hours, with a serum half-life of about 12 hours. Serum levels of 2 to 4 mg/L correlate with clinical efficacy.

c. Special precautions/toxicities: Precipitation or worsening of heart failure (HF) may occur due to its negative inotropic effect; other possible serious reactions are proarrhythmia (similar to quinidine), agranulocytosis, and thrombocytopenia. Common reactions are vagolytic (dry mouth, constipation, urinary retention, exacerbation of glaucoma, abdominal pain). Hypoglycemia and masking of hypoglycemic symptoms (mainly if renal dysfunction is present), as well as triggering of myasthenia crises in patients with this condition, have been reported.

B. Class Ib Drugs

1. Lidocaine

Lidocaine suppresses the electrical activity of depolarized, arrhythmogenic tissue while minimally interfering with the electrical activity of normal tissues. It is effective in the treatment of ventricular tachyarrhythmias but it is no longer the drug of choice, as other agents have proved superior in terminating VT. Its routine prophylactic use in patients with MI has been largely abandoned.

a. Dose: IV bolus of 1.0 to 1.5mg/kg (max 300 mg in 1 hour, total of 3 mg/kg), with immediate maintenance at 1 to 4 mg/min (maintenance should be reduced by 50% in patients with HF, shock, or hepatic dysfunction or who are older than 70 years). Intramuscular or endotracheal administration is feasible, at higher doses.

b. Pharmacokinetics: Lidocaine half-life is 1.5 to 2.0 hours, but it is prolonged with HF, shock, or hepatic dysfunction or in the elderly. Adverse reactions can occur even at therapeutic levels (2–6 mg/L).

c. Special precautions/toxicities: Adverse effects include hypotension (with large doses), central nervous system toxicity (seizures, confusion, stupor, respiratory arrest), bradycardia, and worsening of arrhythmias—for example, acceleration of ventricular response in patients with AF/AFL.

2. Mexiletine

Pharmacologically, mexiletine is similar to lidocaine, orally active but with a less potent antiarrhythmic activity. Indicated only in the treatment of life-threatening ventricular arrhythmias, its initiation should be carried out in the hospital. Because it does not prolong depolarization or repolarization, theoretically it may be useful in the treatment of arrhythmias associated with a long QT interval. It could also be of benefit in the treatment of automatic VT.

a. Dose: When rapid control is needed, an initial loading dose of 400 mg may be administered, followed by 200 mg q8h, which can then be transferred to a q12h dosage if response is adequate. Maximum doses of 400 mg q8h can be used if the drug is well tolerated, allowing 2 to 3 days between dose adjustments.

b. Pharmacokinetics: Onset of therapeutic effect is usually observed within 30 minutes to 2 hours of the first dose. Plasma half-life is 10 to 12 hours, increased in the presence of hepatic dysfunction to > 20 hours. Conditions that increase gastric emptying time—for example, acute MI—reduce its absorption rate.

c. Special precautions/toxicities: Mexiletine has a synergistic effect when used together with class Ia, lidocaine, or class III agents, allowing a lesser dose. Its proarrhythmic effects, mainly TdP, are less common than with type Ia or III drugs. Gastrointestinal manifestations can occur, usually nausea and vomiting. Central nervous system toxicity can include tremor, dizziness, blurred vision, and changes in level of consciousness.

3. Tocainide

Tocainide is a primary amine analog of lidocaine, with similar EP properties, indicated in the treatment of documented life-threatening ventricular arrhythmias. It has been used alone or combined with class Ia drugs. Tocainide has not been shown to be effective in preventing arrhythmia recurrence.

a. Dose: Doses must be individualized on the basis of antiarrhythmic response and tolerance, both of which are dose-related. The recommended initial dose is 400 mg q8h; loss of arrhythmia control prior to next dose suggests use of a shorter interval and/or a dose increase. The usual adult dose is 1200 to 1800 mg/day divided q8h.

b. Pharmacokinetics and metabolism: Bioavailability after oral administration approaches 100%, with peak plasma concentration within 0.5 to 2.0 hours. The average half-life is approximately 15 hours. It has no active cardiometabolites. Therapeutic plasma range is 4 to 10 mg/L.

c. Special precautions/toxicities: Gastrointestinal and neurologic symptoms are similar to those associated with mexiletine use. Agranulocytosis, bone marrow depression, thrombocytopenia, leucopenia, and hypoplastic anemia have been found in 0.18% of patients. It is recommended to monitor the complete blood count weekly for the first 3 months of therapy, and frequently thereafter. Pulmonary disorders (fibrosis, interstitial pneumonitis, fibrosing alveolitis, pulmonary edema, pneumonia) have been reported in 0.11% of patients, usually the seriously ill, appearing following 3 to 18 weeks of therapy.

4. Phenytoin

Phenytoin is used primarily to treat digitalis-induced ventricular and supraventricular arrhythmias, due to its efficacy in stabilizing the membrane potential and reducing hyperexcitability.

a. Dose: The IV loading dose is 100 to 250 mg given over 5 to 10 minutes (maximum 50 mg/min) (alternate regime is 3–5 mg/kg IV over 10 min), with doses of 100 mg q5min until arrhythmia is suppressed, hypotension or symptoms of central nervous system depression ensue, to a total of 1 g.

b. Special precautions/toxicities: Continuous infusion is not recommended. Phenytoin should not be administered with glucose solutions (it precipitates the drug in the IV line). Frequent monitoring of ECG and BP is required. Central nervous system depression is a common side effect, as well as local venous irritation, hypotension, and heart block. Other serious reactions include hematologic disorders, tissue necrosis, Stevens Johnson syndrome, and purple glove syndrome.

C. Class Ic Drugs

1. Flecainide

Flecainide is an effective class Ic agent in the treatment and prophylaxis of supraventricular arrhythmias, including AF and AFL (flecainide itself is more effective in AF than in AFL). Together with other class I agents, it is contraindicated in postinfarction patients.

a. Dose: 50 to 100 mg PO q12h, up to 200 mg q12h at 50 mg bid increments every fourth day. Doses should be reduced if QRS prolongation is excessive.

b. Pharmacokinetics and metabolism: Flecainide absorption is nearly complete, with peak plasma concentration approximately 3 hours after a single dose; it reaches steady plasma concentration in 3 to 5 days with multiple doses. Flecainide is eliminated by hepatic biotransformation, with a half-life of 12 to 27 hours, increased in the presence of renal failure or low cardiac output. Therapeutic plasma concentration is 0.2 to 1.0 mg/L.

c. Special precautions/toxicities: Proarrhythmia (incessant, monomorphic VT) has been observed in 5% to 15 % of patients with ventricular arrhythmias treated with flecainide; aggravation of preexistent arrhythmia may also occur, all with high risk/incidence of SCD. Twenty percent increases in PR and QRS intervals are commonly observed; the dose should be decreased or the medication stopped if the PR increases to > 300 milliseconds, or the QRS prolongs by more than 50%, or if bundle branch block (BBB) or advanced AV block occur. Flecainide could also provoke worsening of HF, or induce sick sinus syndrome (SSS). Atrial and ventricular pacing thresholds often increase, and testing should be performed after the drug is started. Less severe reactions include blurred vision, dizziness, insomnia, paresthesias, nervousness, tinnitus, headache, and constipation.

2. Propafenone

In patients without SHD, propafenone is indicated in the treatment of paroxysmal AF/AFL, paroxysmal SVT, and documented sustained VT. To avoid an increase in ventricular rate in patients with AF or AFL, concomitant treatment with drugs that prolong AVN refractoriness is recommended. Because of proarrhythmia, its use in patients with less severe ventricular arrhythmias is not recommended, even if patients are symptomatic. It has a β-adrenergic-blocking potential (per mg) about 1/40 that of propranolol in humans.

a. Dose: The recommended initial dose is 150 mg PO q8h, which can be increased to 300 mg q8h at a minimum of 3- to 4-day intervals.

b. Pharmacokinetics and metabolism: Peak plasma levels occur approximately 3.5 hours after administration (oral absorption is nearly complete). Bioavailability is variable (5%–50%), increasing with food and dosage (e.g., a 3-fold increase in dose results in a 10-fold increase in plasma concentration), but the relationship between serum level and therapeutic efficacy is poor. In patients with hepatic dysfunction, propafenone clearance is reduced and the elimination half-life is increased. Therapeutic concentration range is 0.2 to 1.5 mg/L.

c. Special precautions/toxicities: In patients with hepatic insufficiency, the dose should be reduced by 20% to 30%; in cases of renal dysfunction, caution must be exercised as the parent drug and its metabolites are eliminated in the urine. Proarrhythmia risk is similar to that seen with flecainide (TdP, other ventricular arrhythmias); other serious reactions are HF, AV block, agranulocytosis, and exacerbation of myasthenia gravis. Bronchospasm may occur as a result of the β-blocking effect. Common reactions include dizziness, taste changes, nausea, dyspnea, and blurred vision. Propafenone increases the effects of warfarin and β-adrenergic antagonists, as well as the plasma concentration of digoxin.

3. Moricizine

Moricizine is a phenothiazine derivative, sharing EP effects of all class I agents. It prolongs the PR and QRS intervals, as well as the atrium-His (AH) and His-ventricle (HV) conduction times. It has minimal effects on ventricular hemodynamics. Useful in the suppression of life-threatening ventricular arrhythmias, it has a considerable risk of proarrhythmia if used post-MI. Additionally, it does not prevent recurrence of sustained ventricular arrhythmias, nor does it improve survival.

a. Dose: 200 to 300 mg PO q8h, with a maximum of 900 mg/day.

b. Pharmacokinetics and metabolism: Moricizine has several hepatic metabolites, with a half-life of 1.5 to 3.5 hours.

c. Special precautions/toxicities: Serious reactions are arrhythmias, ECG/conduction abnormalities, HF, and cardiac arrest.

D. Class II Drugs

β-blockers block the action of endogenous catecholamines on β-adrenergic receptors, of which 2 main types have been classically described: β_1-receptors are mostly cardiac and renal, while β_2-receptors are found mainly in bronchial and vascular smooth muscle. β_3-receptors have recently been described: they are located primarily in fat cells, but evidence of their presence in the heart is increasing (the cardiac role of β_3-receptors is still unclear). β-blockers with higher affinity for β_1-receptors will have a more selective action on the heart, with inhibitory effects on the SANs and AVNs, and on myocardial contraction (while decreasing the risk of noncardiac complications due to nonselective interactions). Apart from their cardioselectivity, β-blockers can have added properties: some induce direct vasodilatation (nebivolol, carvedilol), while others also have α-adrenergic-blocking actions (labetalol). β-blockers such as acebutolol and pindolol exhibit intrinsic sympathomimetic activity (which stimulates β-adrenoceptors), of potential benefit in patients with excessive bradycardia.

The antiarrhythmic effects of β-blockers have several potential mechanisms, with all of them opposing the proarrhythmic effects of increased cAMP and Ca^{2+}-dependent triggered arrhythmias. Their efficacy as anti-ischemic agents decreases significantly the incidence of arrhythmias and arrhythmic death in clinical conditions involving adrenergic arrhythmogenesis, such as CAD and some forms of LQTS (type 1 and possibly type 2). By blocking Na^+-currents at high concentrations, drugs such as propranolol can slow phase 0 upstroke of the AP and suppress ectopic beats by a class I antiarrhythmic action. Sotalol, a β-blocker with class III properties, prolongs the APD (see following for details).

Table 2.2 lists the most commonly available β-blockers and their main properties. β-blockers prevent or terminate arrhythmias associated with excess cardiac sympathetic stimulation (e.g., exercise-induced arrhythmias, arrhythmias due to thyrotoxicosis) and slow the ventricular rate in patients with AF/AFL who are not adequately controlled at rest and during exercise with digitalis. These agents are effective in both atrial and ventricular arrhythmias in patients with mitral valve prolapse, and in ventricular arrhythmias associated with LQTS, decreasing the risk of syncope and sudden death. They are also used to control tachyarrhythmias due to digitalis intoxication when they persist following discontinuation of digitalis and the correction of electrolyte abnormalities. Esmolol is used for immediate control of the ventricular response of AF/AFL or other SVTs.

Special precautions/toxicities of class II drugs are common side effects related to β-blockade, including bradycardias, AV conduction delays, and arterial hypotension. They are contraindicated in cases of advanced AV block, hypotension, acute

Table 2.2. Selected β-blockers

Drug	Additional Activity	Plasma Half-life	Elimination	Standard Dosing
Cardioselective				
Acebutolol	Intrinsic sympathomimetic activity	8–13 h	Hepatic, renal	400–1200 mg/day bid
Atenolol		6–7 h	Renal	PO: 50–200 mg/day od IV: 5 mg slow
Bisoprolol		9–12 h	Hepatic, renal	2.5–10 mg/day od (20 mg rarely)
Metoprolol		3–7 h	Hepatic	PO: 50–200 mg/day bid IV: 5 mg q5min, up to 15 mg
Esmolol		2–9 min		IV: 500 mcg/kg/min x 1 min, followed by 50 mcg/kg/min for 4 min; infusion 100 mcg/kg/min prn
Noncardioselective				
Propranolol		1–6 h	Hepatic	PO: 10–40 mg 3–4 times/day IV: 0.1 mg/kg
Nadolol		20–24 h	Hepatic	40–80 mg/day od (up to 320 mg/day)
Carvedilol	Vasodilator, α-blockade	6 h	Renal	Up to 25 mg bid

HF, and lung disease associated with bronchospasm. Care should be exercised in patients with AF/AFL and known WPW syndrome.

E. Class III Drugs

1. Amiodarone

Amiodarone is structurally related to thyroxine. It has properties of all drug classes due to a combination of antiadrenergic (α and β), Ca^{2+}, K^{+}, and Na^{+}-channel-blocking properties. After IV administration, its major EP effect is increased refractoriness and slowing of AV nodal conduction, with no effect on APD. After prolonged oral administration, its major EP effect is increased refractoriness in all cardiac tissues. Amiodarone is effective in the treatment of VT/VF, AF/AFL, AV nodal reentrant tachycardia (AVNRT), AV reentrant tachycardia (AVRT), and multifocal atrial tachycardia (MAT), and in controlling ventricular response in SVTs.

a. Dose: Several oral loading regimes exist, a common one being 400 mg PO bid x 1 week, then 400 mg PO od x 1 week, then 200 mg PO od (usual maximum maintenance dose is 400 mg PO od). IV loading is often given as 150 mg (3 mL) in 100 mL D5W over 10 min, continuing with 900 mg (18 mL) in 500 mL D5W (1 mg/min for 6 hours, then 0.5 mg/min over 18 hours), avoiding a total dose for the first 24 hours above 2.1 g as the risk of hypotension increases. In fact, because of the relatively slow onset of antiarrhythmic actions and slow elimination, continuous IV maintenance infusions are probably unnecessary. In case of breakthrough arrhythmias, supplemental 150-mg IV doses may be administered.

b. Pharmacokinetics and metabolism: When amiodarone is given orally it has a slow onset of action, with full therapeutic effect taking from 1 week to 3 months; these effects may persist for 7 to 50 days after stopping the medication. It has very high fat solubility and deposits in tissues for long periods. There is a poor correlation between plasma concentration and antiarrhythmic efficacy, partly because of potent activity of the major metabolite desethylamiodarone. Plasma concentration may be detectable up to 9 months after discontinuing the drug.

c. Special precautions/toxicities: Adverse effects are partially dose-related, requiring discontinuation in up to 5% to 10% of patients per year. Among the most important adverse effects are:

- **Lung:** Pulmonary toxicity is a serious reaction, with patients complaining of dry cough, dyspnea, and wheezing associated with pulmonary infiltrates on chest x-ray. If detected early it can be reversible, but undetected it may progress to respiratory failure and death in 10% of cases. Patients on amiodarone should have baseline chest x-rays and pulmonary function tests, followed by repeat history, physical exam, and x-rays every 3 to 6 months. Any new respiratory symptom should suggest the possibility of pulmonary toxicity, and the patient should be evaluated accordingly.
- **Gastrointestinal:** Nausea, vomiting, constipation, and anorexia occur in up to 25% of patients, mainly during the loading phase.
- **Neurologic:** Symptoms are common (up to 20%–40% of patients) and include tremor, involuntary movements, poor coordination, and peripheral neuropathy. The symptoms may respond to reduction in dose or discontinuation of the drug.

- **Ophthalmic:** Corneal microdeposits, usually asymptomatic, develop in almost all patients. Amiodarone can also induce optic neuropathy/neuritis (in some cases progressing to blindness), papilledema, corneal degeneration, photosensitivity, macular degeneration, and lens opacities. Regular ophthalmic examination is recommended.
- **Cardiac:** Exacerbation of arrhythmia is a rare serious reaction; uncommonly, HF may develop, as well as bradycardia and cardiac conduction abnormalities that usually respond to drug discontinuation but may require permanent pacing for control. ECG effects include increased PR, QRS, and QT intervals. Decreasing the rate of infusion can minimize arterial hypotension and bradycardia.
- **Skin:** Photosensitivity is a common adverse reaction, and protection is possible with the use of sun-barrier creams and protective clothing. A blue-gray skin discoloration, of cosmetic significance only, may develop in sun-exposed areas.
- **Thyroid:** Thyroid dysfunction occurs in 2% to 5% of patients per year. Hyperthyroidism may result in thyrotoxicosis and arrhythmia aggravation, and amiodarone should probably be removed if thyrotoxicosis develops. If hypothyroidism develops, it can be managed by dose reduction and concomitant use of levothyroxine. Thyroid function should be monitored yearly.

Amiodarone also has important drug interactions: when used together with digoxin, it results in an increase in the serum digoxin concentration, which may reach toxic levels. Potentiating of warfarin action by slowing its elimination is almost always seen, and can result in serious bleeding (the dose of warfarin should be reduced by one-third to one-half). Increased steady-state levels of quinidine, procainamide, and phenytoin have been reported.

2. Bretylium

Bretylium has direct EP effects, prolonging APD and refractoriness in Purkinje fibers and ventricular myocardium, and increasing the VF threshold; it also has adrenergic neuronal blocking effects. A synergistic effect is noted when used with lidocaine. Bretylium is now rarely used, if ever, because of significant side effects. It may be effective in refractory VT/VF not responding to electric cardioversion (ECV) and first-line drugs.

a. Dose: 5 mg/kg in a rapid IV bolus; if needed, may be repeated at up to 10 mg/kg q5–20 min × 2 (maximum total dose of 35 mg/kg).

b. Pharmacokinetics: Onset of action is rapid, with serum half-life of 4 to 17 hours. Its adrenergic effects have a biphasic response: hypertension and tachycardia in the first 20 minutes after the first dose, followed by hypotension (peaking at 60 min).

c. Special precautions/toxicities: Bretylium may exacerbate digitalis toxicity. Postural hypotension is seen in 60% of patients; other effects (hypertension, tachycardia, nausea, vomiting) are initial and transient. Proarrhythmia is also possible. As a medication, it is being phased out and many hospitals do not stock it anymore.

3. Sotalol

Sotalol is a mixed class II and III agent, having all the properties of β-blockers as well as APD-prolonging drugs. Sotalol may be useful in preventing recurrences of ventricular tachyarrhythmia. It is also often used for maintenance of sinus rhythm after conversion of AF.

a. Dose: 80 mg PO bid, up to 240 to 320 mg daily at 2- to 3-day increments. The drug should be stopped if the QTc exceeds 480 milliseconds and the dose decreased for lesser degrees of excess QT-prolongation.

b. Pharmacokinetics: After oral administration, peak plasma concentrations are reached in 2.5 to 4.0 hours, and steady-state plasma concentrations are attained within 2 to 3 days. Sotalol does not bind to plasma proteins and is not significantly metabolized, being excreted unchanged in the urine. If the creatinine clearance is 30 to 59 mL/min, the daily dose should be reduced by 50% (e.g., by extending the dose interval to 24 hours); if it is 10 to 29 mL/min, the dose should be decreased by about two-thirds to three-fourths.

c. Special precautions/toxicities: The most important adverse effects are TdP (4%) and other serious new ventricular arrhythmias (1%) in high-risk patients (e.g., with history of VT/VF); the risk for other less serious ventricular arrhythmias and SVTs is 1.0% to 1.4%. TdP appears to be dose-dependent and related to the QT interval, worsened by bradycardia and hypokalemia. Proarrhythmic events most often occur within 7 days of initiating therapy or of an increase in dose. Discontinuation because of side effects is high, at 17% of all patients in clinical trials. The most common reactions are fatigue, bradycardia, dyspnea, proarrhythmia, asthenia, and dizziness. Sotalol should not be used in patients with sinus bradycardia or high-degree AV block, and it can trigger episodes of HF and bronchospasm. Concomitant use with medications that prolong the QT should be avoided.

4. Ibutilide

Ibutilide prolongs the atrial and ventricular refractory period (RP) by blocking I_{Kr}. It does not have an apparent effect in sinus node function and the conducting system, and does not affect the BP. Ibutilide is approved for conversion of AF/AFL, with a 43% conversion rate for AF and a 70% for AFL (after a total dose of 2 mg).

a. Dose: 1 mg IV over 10 minutes (0.01mg/kg if < 60 kg); if necessary, a second dose can be repeated 10 minutes after the first one.

b. Pharmacokinetics: Onset of action is almost immediate, with very high distribution and rapid clearance. It is metabolized in the liver and its elimination half-life is 6 hours.

c. Special precautions/toxicities: Serious proarrhythmia is the most significant risk: TdP requiring cardioversion has been documented in 1.7% of patients, nonsustained TdP in 2.7%, and nonsustained SVT in 4.9%. Many episodes occur after the drug is stopped, and the patient should have continuous ECG monitoring for at least 6 hours postinfusion. Before ibutilide administration it is necessary to verify that K^+ and Mg^+ levels are normal, and that the QTc is ≤ 440 milliseconds. Its use should be avoided in patients with congestive HF and LV dysfunction, as the incidence of proarrhythmia is higher.

5. Dofetilide

Dofetilide also blocks I_{Kr}, which results in delayed repolarization. It has no effect on sinus node function and conduction velocities (PR and QRS width remain unchanged), and it does not affect myocardial contractility or BP. Dofetilide is approved for maintenance of sinus rhythm in patients with AF/AFL; even though it is useful in stopping reentrant VT, it does not have formal approval in this setting. Dofetilide also decreases the defibrillation threshold, potentially useful in patients with VT/VF receiving an ICD. Dofetilide is contraindicated in patients with baseline QT > 440 milliseconds (500 milliseconds in the presence of intraventricular conduction abnormalities) or creatinine clearance < 20 mL/min.

a. Dose: Dofetilide should be initiated in hospital, under continuous ECG monitoring for a minimum of 3 days. Doses should be individualized according to renal function and QTc; if both are normal (QT < 440 milliseconds, CrCl > 60 mL/min), the starting dose is 500 mcg PO q12h; with CrCl of 40–60 mL/min, the dose should be reduced to 250 mcg q12h, and to 125 mcg q12 h if CrCl is 20–40 mL/min. The QTc should be checked 2 to 3 hours after the first dose; if QTc increases > 15% over baseline or > 480 milliseconds, the next dose should be decreased by 50%. Then, QTc should be checked q2–3h after each subsequent dose, and if it exceeds 480 milliseconds after a dose reduction, dofetilide should be discontinued.

b. Pharmacokinetics: Oral bioavailability is > 90%, with maximal plasma concentrations in 2 to 3 hours. The terminal half-life is approximately 10 hours and steady-state concentrations are attained within 2 to 3 days. Dofetilide is excreted mainly in the urine.

c. Special precautions/toxicities: TdP is the most significant side effect, with incidence in clinical trials of 3.3% in HF patients and 0.9% in patients with recent MI; TdP is less likely when the dosage is adjusted for renal function and QTc. Dofetilide should not be used with medications known to increase the QT.

6. Dronedarone

Dronedarone is a benzofuran derivative related to amiodarone, with recent market approval. The removal of iodine from the amiodarone structure and an increase in polarity accounts for dronedarone's lack of noncardiac effects. Dronedarone promotes maintenance of sinus rhythm after conversion of AF/AFL but is less effective than amiodarone. Dronedarone is contraindicated in patients with class IV HF, or class II–III HF with a recent decompensation requiring hospitalization or referral to a specialized HF clinic.

a. Dose: 400 mg PO bid.

b. Pharmacokinetics and metabolism: Dronedarone is less lipophilic than amiodarone, has a much smaller volume of distribution, and its elimination half-life is about 24 hours.

c. Special precautions/toxicities: The most common reactions are gastrointestinal disorders (diarrhea, nausea, abdominal pain, vomiting), rash, and creatinine elevation. Acute liver failure has been reported. Patients may also develop bradycardia and QT-interval prolongation.

F. Class IV Drugs

1. Verapamil

Verapamil is a nondihydropyridine slow CCB that prolongs the effective RP of the AVN, increasing its refractoriness and slowing the AV conduction, thus making it effective in terminating AV nodal reentrant arrhythmias and controlling the ventricular response in patients with AF/AFL. Even though it is effective in verapamil-sensitive VT, verapamil can also accelerate the rate in other types of VT; therefore, it should be used in narrow-complex tachycardias only when there is certainty of a supraventricular origin. The sinus rate is usually not affected, but in patients with SSS, verapamil can lead to sinus arrest/sinoatrial block via interference with sinus node impulse generation. Verapamil may shorten the antegrade effective RP of an accessory pathway, resulting in acceleration of ventricular rate and possible degeneration to VF. Because of its effects on smooth muscle cells, verapamil produces a significant decrease in BP.

a. Dose: The initial IV dose is 2.5 to 5.0 mg over 2 minutes; if needed, repeated doses of 5 to 10 mg can be given q15–30min to a total of 20 mg. Oral maintenance dosing ranges from 240 to 480 mg daily (sustained release presentation).

b. Pharmacokinetics: With IV use, verapamil has a very rapid onset of action, 1 to 2 minutes, with peak effect in 10 to 15 minutes. Orally, peak plasma concentrations are reached in 1 to 2 hours; its mean elimination half-life is 4 to 12 hours. Verapamil has several hepatic metabolites.

c. Special precautions/toxicities: Bradycardia, high-degree AV block, and hypotension could occur, and the dose should be reduced or the drug discontinued. Verapamil should not be given to patients with severe LV dysfunction or HF (because of its negative inotropic effects), or to patients with HF receiving a β-blocker. Constipation develops in 7% of patients. Verapamil should not be used in patients with known accessory pathway.

2. Diltiazem

Diltiazem has similar EP effects and clinical use to verapamil, but its shorter half-life and less influence on myocardial contractility make it more attractive for use in severe LV dysfunction. Diltiazem has frequency (use) dependent effects on AVN conduction, allowing reductions in HR during tachycardias involving the AVN with little or no effect on normal AVN conduction (at normal HR).

a. Dose: Initial IV dose is 0.25 mg/kg (20 mg for the average patient), followed by a second dose when necessary, after 15 minutes, at 0.35 mg/kg (average 25 mg); maintenance is 5 to 15 mg/h, for up to 24 hours. The usual oral starting dose is 120 to 240 mg/day.

b. Pharmacokinetics: The plasma half-life is 3 to 4 hours after a single IV injection, with a nonlinear distribution if continuous administration is used (increasing half-life by up to 50%). After oral administration, it undergoes extensive hepatic metabolism and conjugation, with a half-life of about 6 hours. Sustained-release preparations allow once-daily dosing.

c. Special precautions/toxicities: Side effects are similar to those of verapamil, with hypotension being the most common, generally mild and transient. Its concomitant use with β-blockers has been well tolerated in clinical trials, but caution is needed to avoid excessive sinus bradycardia or AV block. Despite its negative inotropic effects, no consistent negative effect on contractility has been demonstrated; because the experience is limited, its use in patients with impaired ventricular function should be cautious.

G. Class V Drugs

1. Adenosine

Adenosine is an endogenous nucleoside that acts to slow conduction through the AVN, prolonging its refractoriness. It is indicated for conversion of SVT or suspected SVT with aberrancy. It should not be used in preexcitated SVTs due to a high risk of rapid conduction through the accessory pathway.

a. Dose: 6 mg IV initial dose. Additional doses of 12 mg IV q1–2min x 2 may be given. Because of its extremely short half-life, IV access should be in a large vein above the diaphragm if possible (e.g., antecubital vein); giving all doses in a rapid push over 1 to 3 seconds, followed by a 10- to 30-cc flush of normal saline.

b. Pharmacokinetics: Adenosine has a very short half-life because of rapid destruction by adenosine deaminase, <10 seconds, not affected by hepatic or renal failure. Continuous IV infusion is not effective and should not be used.

c. Special precautions/toxicities: The initial dose should be reduced to 3 mg in patients taking dipyridamole or carbamazepine, in those with transplanted hearts, or if given by central venous access. Larger doses may be required in patients taking theophylline, caffeine, or theobromine. Common reactions (facial flushing, chest pain, dyspnea) are usually brief. Advanced or complete AV block are also fairly common and of short duration; however, it could occasionally lead to prolonged asystole, VF, or bronchospasm. If needed, effects can be antagonized with methylxanthines.

2. Digitalis

By inhibiting the sodium-potassium ATPase in nerve cells and augmenting vagal tone to the heart, digoxin prolongs AVN conduction time and increases AVN refractoriness. Digoxin also has a significant positive inotropic effect. Digoxin is mainly indicated to control the resting ventricular rate in AF/AFL.

a. Dose: The usual total loading dose is 0.75 to 1.25 mg PO or 8 to 12 mcg/kg IV in divided doses (50% of the total dose given initially and then 25% of the total dose in each of 2 subsequent doses 8–12 hours apart) while monitoring the ventricular response. Maintenance dosing is 0.125 to 0.25 mg od.

b. Pharmacokinetics: After oral administration, peak serum concentrations occur in 1 to 3 hours (with enhanced absorption if capsules are used instead of tablets), with a time to peak effect of 2 to 6 hours. The therapeutic range is 0.8 to 2.0 ng/mL, but digoxin may produce clinical benefits at lower concentrations. Digoxin has a half-life of 1.5 to 2 days. Excretion is mixed, mainly renal in normal individuals, with a half-life of 36 hrs (7 days in anephric individuals).

c. Special precautions/toxicities: Adverse reactions of digoxin are dose-dependent, with a variety of rhythm disturbances (varied degrees of AV block, AT with block, AV dissociation, accelerated junctional rhythm, PVCs, VT/VF) seen at toxic doses. Therapeutic doses of digoxin may cause heart block in patients with preexisting SA or AV conduction disorders. Digoxin produces PR prolongation and ST segment depression, which are not signs of toxicity. Other toxic reactions are anorexia, nausea, vomiting, diarrhea, blurred or yellow vision (an early sign of toxicity), headache, weakness, and mental disturbances (anxiety, depression, delirium, hallucinations). K^+-depleting diuretics are a contributing factor to digitalis toxicity because hypokalemia enhances the drug's effects. Quinidine, verapamil, amiodarone, propafenone, indomethacin, itraconazole, alprazolam, and spironolactone raise the serum digoxin concentration, potentially causing toxicity. It is suggested to reduce the dose of digoxin for 1 to 2 days prior to ECV of AF to avoid the induction of ventricular arrhythmias, or to use the lowest possible energy level.

III. PROARRHYTHMIA

Proarrhythmia is the provocation of a new arrhythmia or exacerbation of a preexisting arrhythmia as a result of AAD therapy. Proarrhythmia may occur early or late during the course of therapy. Proarrhythmia is not predictable and may occur even when AAD levels are within therapeutic limits. Though it can be the direct result of a drug's EP effects on conduction velocity, refractoriness, and automaticity, it may also be the result of metabolic abnormalities, changes in autonomic state, or drug/drug interactions. Class Ic drugs have the highest proarrhythmic risk in patients with SHD. Other known risk factors are long QT interval, preexisting conduction disturbances, sinus node dysfunction (SND), older patients, and the presence of HF.

Recognized types of proarrhythmia are:

- **TdP:** Occurs in 1% to 8% of patients exposed to AAD that increase the QT, typically sotalol, quinidine, dofetilide, and ibutilide. Known risk factors are female sex, underlying heart disease, hypokalemia, and hypomagnesemia. Management includes appropriate recognition, withdrawal of the offending agent, empiric administration of IV Mg^+ (regardless of serum levels), correction of serum K^+ to 4.5 to 5.0 mEq/L, and measures to increase HR (isoproterenol or pacing) if necessary.
- **Reentry:** Reentry is mainly induced by Na^+ channel blockers. Its incidence is higher in cases of postinfarct VT, due to slow conduction in the border zone.
- **Others:** These include acceleration of the tachycardia rate (e.g., AFL with rapid AVN conduction, when class I AADs are not administered combined with an AVN-blocking agent); conversion of AF into AFL; synergistic action of Na^+ channel block and recurrent myocardial ischemia to provoke

VT/VT in patients with recent infarction; prolonged QT intervals, SND, and new or worsened AV conduction abnormalities; increase of defibrillation or pacing thresholds.

SUGGESTED READING

Opie LH, Gersh BJ, eds. *Drugs for the Heart.* 7th ed. Philadelphia: Saunders. 2009.

PART 2

Disorders of Impulse Formation and Conduction

Chapter 3

Sinoatrial Node Disorders

Miguel A. Barrero Garcia and Mario Talajic

I. GENERAL PRINCIPLES

The sinoatrial node (SAN) is a conglomerate of specialized pacemaker cells (P cells) located at the junction of the high right atrium (RA) and the superior vena cava; being commonly the most rapid pacemaker it is responsible for the generation of the normal cardiac rhythm (Table 3.1).

Located around the central P cells, perinodal cells (T cells) allow the transmission of the electrical impulses from the sinus node to the RA. Conditions that affect the functioning of the sinus node or its SA extension lead to disorders of impulse formation or conduction, broadly referred to as sinus node dysfunction (SND) or sick sinus syndrome (SSS) (Table 3.2).

SND has multiple clinical presentations. Most cases of primary SA dysfunction will require pharmacological therapy or even permanent pacing whereas secondary cases of SND, which is often precipitated by a drug, are potentially reversible by controlling the offending agent.

SSS is the classic example of a malfunctioning SAN, but there are other abnormalities in sinus impulse formation/propagation, often asymptomatic and of benign nature.

II. SPECIFIC ENTITIES

A. Sinus Arrhythmia

Sinus arrhythmia is characterized by a sinus P-wave morphology (PWM)/axis, with a gradual phasic change in PP interval of > 10% (or 120 milliseconds); the P morphology may change when the sinus rate alters. There are 2 forms of sinus arrhythmia: in the "respiratory" form, the R-R interval shortens during inspiration and lengthens during expiration (breath-holding eliminates the variation); in the "non-respiratory" form, the same phasic variation is seen in the R-R interval but is not related to respiration. This latter form of sinus arrhythmia occurs in elderly patients, in those with digoxin overdose, and in patients with increased intracranial pressure.

Table 3.1. Electrocardiographic Features of Normal Sinus Rhythm

Parameter	Feature
HR	60–100 bpm
P-wave axis	0 to +90°
PWM	Upright in DII, DIII, and usually aVF; inverted aVR; upright or biphasic V_1 and V_2; upright V_3 to V_6
PR interval duration	120–200 ms
PP interval	Does not vary significantly

Table 3.2. Etiology of Sinus Node Dysfunction

Intrinsic SND	Extrinsic SND
Degenerative (most common)	Drugs (most AADs, β-blockers, CCBs, cimetidine, lithium, phenytoin)
Ischemic	Electrolytes (e.g., hyperkalemia)
Infiltrative (e.g., amyloidosis, tumors)	Endocrine (hypothyroidism)
Inflammatory (e.g., pericarditis, myocarditis)	Inferior myocardial infarction (neural reflex)
Musculoskeletal (Duchenne's, myotonic dystrophy, Friedreich's)	Neurally mediated bradycardia/ hypotension
Collagen vascular disease (lupus, scleroderma)	Others (intracranial hypertension, jaundice)
Postoperative (Mustard, atrial septal defect)	

Respiratory sinus arrhythmia is due to physiologic fluctuations in autonomic tone: "no treatment" is indicated.

B. Wandering Atrial Pacemaker

Wandering atrial pacemaker is an atrial rhythm with at least 3 different PWMs at rates between 50 and 100 bpm (cycle length [CL] 1200–600 milliseconds). It is usually considered benign and no specific therapy is necessary, unless there is symptomatic bradycardia.

C. Sinus Bradycardia

Sinus bradycardia is characterized by a sinus rate < 60 bpm (CL >1000 milliseconds), with a normal PWM and axis. It occurs in normal children and adults but it can also result from exaggerated vagal activity (a HR < 40 bpm during sleep has a prevalence of 24% in healthy people). Conditions leading to sinus bradycardia include increased intracranial pressure, obstructive sleep apnea, acute MI, and drug effect (β-blockers, CCBs, amiodarone).

In many patients sinus bradycardia is asymptomatic but others may present with dyspnea on exertion, fatigue, presyncope/syncope, angina, or HF. The ECG is diagnostic. Treatment of symptomatic sinus bradycardia includes discontinuing responsible medications, use of vagolytic drugs like atropine, temporary pacing (permanent in cases of pathologic bradycardia), and adequate management of underlying conditions.

D. Sinoatrial Exit Block

Sinoatrial exit block is a disorder of impulse propagation. First-degree SA block is determined by measurement, direct or indirect, of the SA conduction time (it cannot be recognized in a surface ECG): it indicates a slowing of impulse exit but 1:1 conduction persists. A third-degree SA block is also very difficult to distinguish clinically from a prolonged sinus arrest/pause, unless there is a direct recording of sinus node activity. The only SA exit block that can be diagnosed clinically is second-degree block (Table 3.3).

Treatment of SA exit block, if symptomatic, is similar to that of sinus bradycardia or SSS.

E. Sinus Arrest/Pause

Sinus arrest is a pause during sinus rhythm, without a P wave, lasting > 2 seconds; the PP interval of the pause is not a multiple of the basic PP interval but longer. It is thought that sinus arrest results from slowing or cessation of spontaneous sinus node automaticity; therefore, it is a disorder of impulse formation. Differentiation from SA exit block requires direct recordings of sinus node discharge. Pauses lasting < 3 seconds are considered benign, but when longer they usually have clinical implications and a comprehensive evaluation of the patient's symptoms is necessary.

Table 3.3. Electrocardiographic Features of Second-Degree Sinoatrial Exit Block

Parameter	Mobitz I SA Exit Block	Mobitz II SA Exit Block
PWM/axis	Normal	Normal
PR interval	Normal and constant	Normal and constant
Pause	No visible sinus P wave	No visible sinus P wave
PP interval	Progressive shortening before the pause (Wenckebach periodicity)	Constant PP before and after the pause
PP of pause vs PP preceding the pause	PP of pause less than twice the PP preceding the pause	Pause is an integral multiple (within 100 ms) of normal PP

Pauses during sleep are diagnosed with increasing frequency: though benign in many cases (prevalence of 4%–10%), an association with SHD or the presence of symptoms warrants appropriate investigation.

F. Ectopic Atrial Rhythm

Ectopic atrial rhythm arises from a site other than the SAN, with a rate between 60 and 100 bpm (CL 1000–600 milliseconds). Though not exactly a sinus node disorder by itself, its presence is an indication of depressed nodal function or atrial disease. Ectopic atrial rhythm is usually benign and no specific therapy is needed.

G. Sinus Tachycardia

Sinus tachycardia is an arrhythmia emanating from the SAN at a rate > 100 bpm (CL < 600 milliseconds), with a gradual onset and termination, and proportional to the level of physical, emotional, pathological, or pharmacological stress. Some common etiologies are physical exercise, fever, anemia, sepsis, hyperthyroidism, anxiety, myocardial ischemia, HF, and caffeine use. In many cases sinus tachycardia is a physiological response; when accompanying conditions like HF, it could either indicate a compensatory response or herald a worsening condition. When sinus tachycardia persists in patients postinfarction, it is a reliable indicator of increased morbidity and high early mortality. Its most significant characteristics are exposed in Table 3.4.

Physiologic sinus tachycardia does not need specific management; when secondary to underlying diseases, treatment is aimed at them.

Table 3.4. Selected Characteristics of Sinus Tachycardia, IST, and POTS

Parameters	Sinus Tachycardia	IST	POTS
Triggers	Stress, sepsis, varied heart conditions	Unknown	Stand-up position
Symptoms	None, rapid heartbeat	Incessant palpitations	Palpitations
Onset/termination	Gradual	Persistent	Gradual
PWM/axis	Sinus	Sinuslike	Sinus
PR interval	Shortens	Shortens	Shortens
Response to vagal maneuvers/ AVN-blocking agents	Slowing without other changes	Slowing, PWM may change slightly	Slowing without other changes

H. Inappropriate Sinus Tachycardia

Inappropriate sinus tachycardia is a rare condition characterized by high intrinsic HR, depressed efferent cardiovagal reflex, and hypersensitivity to β-adrenergic stimulation, all thought to result from a primary sinus node abnormality. It is most commonly seen in patients without heart disease or other cause for sinus tachycardia. Patients usually complain of persistent palpitations, almost incessant, with exercise intolerance and occasional presyncope (see Table 3.4 for further details). Females make up 90% of affected patients, many of them health care workers. There is significant overlap of this syndrome with postural orthostatic tachycardia syndrome (POTS) and chronic fatigue syndrome. Holter testing usually shows a mean HR > 90 bpm, with exercise stress testing (EST) having HR > 130 bpm within the first 90 seconds of the test. During electrophysiological study (EPS), inappropriate sinus tachycardia (IST) is not initiated by atrial stimulation (which serves as differential diagnosis from sinus node reentrant tachycardia [SNRT] and atrial tachycardia [AT]).

Treatment of IST with a β-blocker may be effective, but control is difficult if there is a depressed vagal activity: some of these patients have undergone partial ablation (modification) of the SAN, but this procedure may not eliminate all symptoms and is associated with a high risk of significant damage and subsequent need for permanent pacing.

1. Postural Orthostatic Tachycardia Syndrome

POTS is a disorder of the autonomic nervous system associated with orthostatic intolerance (the presence of symptoms on standing that are relieved by lying down). It is a heterogeneous condition, classified into primary and secondary forms. The primary form is not associated with any other disorder and is further subclassified in 2 groups: the partial dysautonomic form and the hyperadrenergic one.

In the partial dysautonomic form, POTS is characterized by an orthostatic rise in HR > 30 bpm above baseline, or > 120 bpm within the first 10 minutes of a head-up tilt table test (TTT), accompanied by palpitations and no significant (< 20/10 mm Hg) fall in BP. It is most common in young women with normal hearts. The hyperadrenergic form also has orthostatic tachycardia, but patients are usually hypertensive and have high upright serum norepinephrine levels.

The secondary form of POTS occurs due to conditions such as diabetes, amyloidosis, lupus, and sarcoidosis.

In patients with suspected POTS, a detailed history and thorough physical examination are very important, with HR and BP taken while the patient is supine, while the patient is sitting, immediately on standing, and after intervals of 2, 5, and 10 minutes. A TTT is recommended for most patients. Treatment is based on cessation of drugs known to cause orthostatic intolerance, physical reconditioning, adequate

Table 3.5. EP Characteristics of SNRT

- It can be initiated by atrial premature beats and atrial pacing, as well as by ventricular premature beats and ventricular pacing in patients with retrograde VA conduction (which serves as differentiation from intra-atrial reentry)
- Termination with premature atrial stimuli
- Constant intra-atrial conduction time

hydration, and use of compressive stocking, paired if needed with oral fludrocortisone and/or midodrine. The hyperadrenergic form of POTS often responds to clonidine.

2. Sinus Node Reentrant Tachycardia

SNRT results from a reentrant circuit within the SAN, with an activation sequence that resembles that of normal sinus rhythm (NSR). Usually associated with older age and SHD, this arrhythmia is relatively uncommon and symptoms are usually nonsignificant. In the surface ECG, SNRT is indistinguishable from sinus tachycardia, with rates usually between 100 and 150 bpm; clinically, its abrupt onset and termination makes the differentiation. EP features of SNRT are exposed in Table 3.5.

If needed, SNRT episodes can be treated with vagal maneuvers (effective due to the extensive autonomic innervation of the sinus node), adenosine, or verapamil. Catheter ablation may have a role in severely symptomatic patients.

3. Sick Sinus Syndrome

Sick sinus syndrome is a clinical syndrome usually due to chronic SND, with varied clinical manifestations not only due to abnormalities of sinus impulse formation/propagation but also associated intra-atrial and AV conduction abnormalities,

Table 3.6. Clinical Forms of SSS

- Pathologic sinus bradycardia
- Periodic sinus pause/arrest or SA block, with or without appropriate escape rhythms
- AF with slow ventricular response in the absence of AVN- blocking agents
- Failure of NSR to resume after cardioversion or spontaneous termination of atrial tachyarrhythmias
- Tachycardia-bradycardia syndrome
- Paroxysmal AF followed up by sinus tachycardia

failure of subsidiary pacemakers, and supraventricular tachyarrhythmias (Table 3.6). Historically SSS has been diagnosed in the setting of sluggish or absent sinus node pacemaker activity after electric cardioversion (ECV); another typical setting is alternating episodes of bradycardia and supraventricular arrhythmias such as AF (tachycardia-bradycardia syndrome). One frequently associated finding is chronotropic incompetence, which is the inability to increase the HR during exercise, with patients usually not surpassing 70% to 85% of their age-predicted maximum (220 minus age) or not reaching a HR of 120 bpm.

Patients with SSS have an increased risk of cardiovascular events, including thromboembolism; they also have a significant mortality due to noncardiac causes.

Most patients seek consultation due to dizzy spells, palpitations, dyspnea on exertion, presyncope/syncope, or worsening angina. Occasionally symptoms are nonspecific and the ECG features are either not documented or nondiagnostic.

III. DIAGNOSTICS FOR SINOATRIAL NODE DYSFUNCTION

A. Electrocardiogram

An ECG provides confirmatory information when abnormalities are documented (Figure 3.1).

B. Determination of Potential Neural Contributors

1. Carotid sinus massage

Carotid sinus massage is among the first tests done when SSS is suspected; if positive (see Chapter 13 for further information), with reproduction of similar symptoms, no further testing is usually necessary. Occasionally, SSS and carotid sinus hypersensitivity coexist as separate entities.

2. Tilt table testing

TTT is mostly helpful in patients with suspected vagal component, presenting with presyncope/syncope and nondiagnostic ECG.

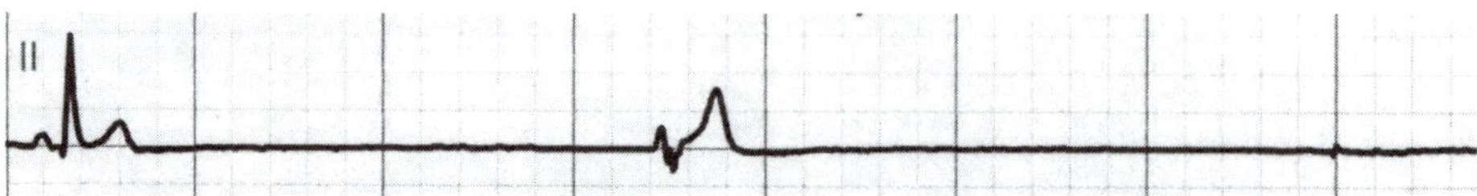

Figure 3.1. SND: several hours after successful ECV for AF there is sinus arrest with no escape rhythm for 3.1 seconds; a nodal escape beat is then followed by asystole (until NSR was restored with IV atropine).

C. Holter Monitoring

Holter monitoring helps to correlate symptoms with electrical disorders, but the short duration of the test (usually 24–48 hours) limits its utility to those patients with symptoms on a regular basis (almost daily).

D. Event Loop Recorders

External event loop recorders have similar indications as Holter testing but the length of testing (2–3 weeks) makes them more appropriate in patients with less frequent symptoms.

E. Exercise Stress Testing

EST helps fundamentally to assess the chronotropic response of the patient to exercise. In patients with excessively rapid deceleration after exercise, it may indicate the need for sensor-triggered rate-adaptive pacing.

F. Assessment of Intrinsic Heart Rate

Intrinsic heart rate (IHR) is the sinus node rate when neural control is eliminated by pharmacological autonomic blockade (IV atropine, 0.04 mg/kg, + IV propranolol, 0.2 mg/kg). The observed IHR is then compared to a predicted IHR (IHRp = 118.1 – [0.57 x age]). In the setting of bradyarrhythmias a normal IHR suggests autonomic imbalance whereas a low IHR points towards intrinsic SND; a high IHR is an indication of IST. This test is usually not indicated clinically.

G. P-wave Signal Averaging

Long, low-amplitude signals early during signal-averaged P waves are characteristic of SSS, with an estimated sensitivity and specificity of 78% and 84%, respectively.

H. Electrophysiological Study

An electrophysiological study (EPS) is not necessary in most patients. EP evaluation of sinus node disorders is relatively specific but lacks sensitivity. Potential indications are summarized in Table 3.7.

Helpful measurements obtained during EPS are:

- **Sinus node recovery time:** The most used and useful method to test for sinus node automaticity, based on the property of cardiac cells to be suppressed by pacing at rapid rates. The time that it takes the sinus node activity to return after the pacing is abruptly stopped (the SNRT) should be

Table 3.7. Suggested Indications for EPS in Suspected SND

Symptomatic patients, no known cause for symptoms and nondiagnostic ECG
No correlation between symptoms and ECG
Symptomatic SSS, with NSR while asymptomatic (if sinus node function is normal, autonomic testing should be carried out: if positive, medical therapy is possible)
SND on usual doses of AADs, β-blockers, CCBs, or digitalis

< 1500 milliseconds to be considered normal. The SNRT can be corrected (CSNRT) by subtracting the sinus CL from the SNRT (normal values < 525–550 milliseconds), or by using the following formula: CSNRT = $\frac{SNRT}{Sinus\ CL}$ x 100 (normal if < 160%). If SSS is strongly suspected clinically but the CSNRT is not positive, the test may be repeated after administration of adenosine or disopyramide to increase its sensitivity.

- **SA conduction time:** The time required for the sinus impulse to activate the atrium; its usefulness is limited by the need to avoid pacing too early or too late into the sinus cycle to avoid interpolated stimulus, local reentry, and/or other mishaps.
- **Total recovery time:** The time from cessation of pacing to return to basic CL (normal if < 5 seconds); it is used to account for variability in cycles when the NSR returns.
- **Sinus node and atrial refractory period (RP):** Sinus node effective RP normally measures 250 to 350 milliseconds (500–550 milliseconds in patients with SSS).
- **Duration of sinus node EGM:** A direct recording of sinus node activity; it normally measures < 120 milliseconds.

IV. THERAPY

Treatment of SND depends basically on the symptomatology and includes removing offending agents in drug-related cases, resolution of ischemia and control of autonomic imbalance, and implantation of a permanent pacemaker in cases of symptomatic bradyarrhythmias, either related with the primary SND or secondary to drugs deemed essential to treat associated disorders (Table 3.8).

Most patients with SND should receive a rate-responsive dual-chamber pacemaker; dual-chamber pacing reduces the incidence of AF but has no effects on stroke or mortality as compared with single-chamber stimulation pacing. If there are no suspected AV conduction abnormalities, and if their risk in the future is considered

Table 3.8. Recommendations for Permanent Pacing in SND

Class I

- SND with documented symptomatic bradycardia/pauses
- Symptomatic chronotropic incompetence
- Symptomatic sinus bradycardia that results from required drug therapy for medical conditions

Class IIa

- SND with HR < 40 bpm when a clear association between symptoms and bradycardia has not been documented
- Syncope of unexplained origin with positive EPS for significant SND

Class IIb

- Patients with chronic HR < 40 bpm while awake, minimally symptomatic

Class III

- Asymptomatic SND
- SND with symptoms suggestive of bradycardia that are clearly documented as not associated with the slow HR
- SND with symptomatic bradycardia due to nonessential drug therapy

low, a single-chamber atrial pacemaker may suffice. For patients with permanent AF or AFL, a single-chamber ventricular pacemaker is indicated.

Once a pacemaker has been implanted, drug therapy type and dosage can be optimized if there is a need to control symptomatic tachyarrhythmias or to treat associated medical conditions. Some patients with drug-refractory atrial tachyarrhythmias may benefit from complete ablation of the AVN for appropriate rate control if their arrhythmia is not amenable for a curative atrial-based ablation.

V. CARDIOLOGY EVALUATION

Cardiology evaluation is recommended when there is a potential indication for EPS. Patients with symptomatic SND (e.g., IST, POTS, or SNRT) should be referred to specialized physicians; the same applies to all cases of SSS requiring nonpharmacological therapy.

SUGGESTED READING

Epstein AE, Dimarco JP, Ellenbogen KA, et al. ACC/AHA/HRS 2008 guidelines for device-based therapy of cardiac rhythm. *J Am Coll Cardiol.* 2008;51:2085-2105.

CHAPTER 4

Atrioventricular Conduction Disorders

Miguel A. Barrero Garcia and Mario Talajic

I. GENERAL PRINCIPLES

The AV node (AVN) has classically been defined as a small nodular mass of specialized cardiac cells, located in the interatrial septum anterior to the coronary sinus, that conducts the normal electrical impulse from the atria to the ventricles (see Figure 1.1). It has the unique property of decremental conduction (the more frequently the AVN is stimulated, the slower it conducts), which prevents rapid conduction to the ventricles in case of accelerated supraventricular rhythms.

The anatomic substrate of the AVN is somewhat unclear: the compact node has nodal inputs in the atrium as well as an extension into the bundle of His; there is also a posterior nodal extension, situated above the insertion of the tricuspid valve beneath the coronary sinus ostium. The functional anatomy is also controversial, with 3 recognized zones: the AN (supranodal), N (nodal), and NH (infranodal) regions, with the slowest conduction attributed to the NH region. Interestingly the action potential (AP) of these regions can be recorded not only from the compact node area but also from a much broader zone. Apart from its rate-dependent properties (recovery, facilitation, and fatigue) the AVN has dual conducting pathways, fast and slow, responsible for reentrant AVN tachyarrhythmias (see Chapter 9 for further details on this subject). Once the cardiac impulse traverses the AVN, it then conducts through the His bundle, bundle branch system, and Purkinje network until it reaches ventricular myocardium. Unlike the decremental properties of the AVN (which has a conduction speed of around 0.05 m/s), the His-Purkinje system conducts impulses rapidly, at about 2 m/s at the bundle of His and the bundle branches and at 4 m/s at the Purkinje fibers.

Impairment of AVN conduction ranges from delayed transmission of atrial impulses to the ventricles (first-degree AV block), to intermittent failure of conduction (second-degree AV block), to complete failure of conduction (third-degree AV block). Patterns of block correlate well with abnormalities of particular AVN regions, but variability does occur.

II. FIRST-DEGREE AV BLOCK

A. General Principles

A first-degree AV block is diagnosed when the PR interval exceeds 200 milliseconds. It is uncommon in young healthy adults but becomes more frequent as people age, being diagnosed in up to 5% of men > 60 years old, and as high as 10% in patients with structural heart disease (SHD).

B. Mechanisms

The site of the block could potentially be located at any level of the AV conduction system, but most often it is found in the compact zone of the AVN (> 85% of patients with narrow QRS); if the QRS is wide, its origin is infranodal in 45% of cases. Occasionally first-degree AV block is the result of a heightened vagal tone, as in highly trained athletes (in these cases, it disappears with exercise). Other etiologies include drug influence (e.g., β-blockers, CCB), degenerative or infiltrative processes affecting the conduction system, ischemia, cardiac surgery, collagen vascular diseases, and infectious disease.

C. Clinical History

First-degree AV block is usually asymptomatic and, if isolated, has no impact on prognosis; rarely, an extremely prolonged PR interval can lead to heart failure (HF) as a result of dyssynchronism in AV contraction, impaired ventricular filling, and reduction in cardiac output. In the setting of acute inferior myocardial infarction (MI) it may herald higher degrees of AV block.

D. Diagnosis

1. Electrocardiogram

Diagnosis by ECG involves a PR > 200 milliseconds; each P is followed by a QRS complex.

2. Laboratory testing

Specific tests are not routinely needed.

3. EP study

An EPS is usually not indicated; the most common finding is 1:1 AV conduction with a long atrium-His (AH) interval.

E. Therapy

No specific management is ordinarily required. In the rather rare instances of symptomatic block, medications that slow AV conduction should be stopped if possible; a permanent pacemaker implant may be considered.

F. Cardiology Evaluation

Cardiac consult is suggested if a significant first-degree AV block (e.g., PR > 300 milliseconds) is associated with HF or syncope.

III. SECOND-DEGREE AV BLOCK

A second-degree AV block is characterized by intermittent failure of conduction of some atrial beats into the ventricles; depending on its presentation and severity it is classified as type I or type II.

A. Type I Second-degree AV Block

1. General principles

The classical type I second-degree AV block (Mobitz I block), with periods of Wenckebach and Luciani, is characterized by gradual PR interval prolongation (Wenckebach sequence) leading to a blocked P wave (Luciani phenomenon), and then reinitiating of the sequence. It usually carries a benign prognosis. Similar to first-degree AV block, this type of block may occur in healthy, well-conditioned people as a physiologic manifestation of high vagal tone (it is observed in 6%–11% of healthy people, especially during sleep). Other causes of Mobitz I second-degree AV block are the same as in first-degree AV block (see previous section).

2. Mechanisms

Type I second-degree AV block is most commonly due to a delay of conduction at the level of the AVN (but not always). If the QRS is narrow the site of block is usually the nodal region of the AVN (mainly if the increment in PR is > 100 milliseconds, or if there is sinus slowing with AV block), the exception being intra-hisian block, which usually has a narrow QRS. When the QRS is wide the origin of the block is likely infranodal (unless there is a preexistent intraventricular conduction defect, like a right bundle branch block [RBBB]). Mobitz I second-degree AV block may also occur physiologically at high heart rate (HR) (especially with pacing) due to increased refractoriness of the AVN, which protects against conducting a fast rhythm to the ventricles.

3. Clinical history

Mobitz I second-degree AV block is typically asymptomatic, benign, and nonprogressive (except in the rare cases with an infranodal origin, which may progress to complete block). Some patients may sense irregularities in heart rhythm.

4. Diagnosis

a. Electrocardiogram: It shows the classic Wenckebach sequence (Figure 4.1): progressive PR lengthening, followed by a blocked P wave; shortening of the PR interval that follows the blocked cycle (when compared to the last conducted PR); gradual R-R shortening prior to the blocked P wave (the increment in PR interval between the first and second beat of a cycle is the largest; in subsequent beats, the increments progressively decrease); the pause after the blocked P is less than twice the PP interval.

The classic Wenckebach pattern usually occurs with ratios 3:2, 4:3, or 5:4; this gives rise to a typical clustering of beats (group beating), with decreasing R-R intervals that tend to repeat.

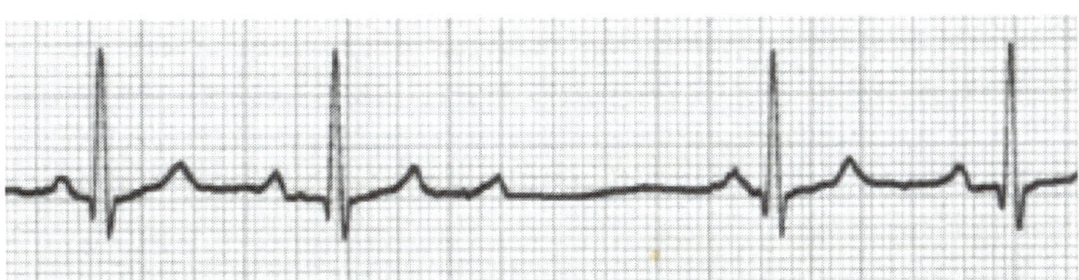

Figure 4.1. Mobitz I second-degree AV block.

Even though the Wenckebach periodicity is the most calling feature of this type of block, it could be atypical when the degree of block is higher than 6:5; however, even in these cases the PR that follows the blocked P wave is shorter than the PR interval before the blocked beat.

b. Laboratory testing: Renal function and drug levels are needed if there is suspicion of electrolyte abnormalities or drug toxicity. Cardiac enzymes are necessary if suspicion of acute coronary syndrome. If clinical evaluation suggests systemic illness, appropriate directed laboratory studies for infection, myxedema, or connective tissue disease should be performed.

c. EP Study: Usually not required, unless you are dealing with a patient with unexplained presyncope/syncope and suspected AV block as the cause of it. In cases of nodal block (70%–75% of cases), the AH interval progressively lengthens (corresponding to the Wenckebach sequence in the ECG) until an atrial stimulus is not followed by activation of the His potential. In cases of infranodal block, there is a progressive prolongation of the HV interval (usually long from the beginning of the cycle) until a His depolarization is not followed by ventricular activation: this finding may guide a decision for pacemaker therapy.

5. Therapy

Usually no specific treatment is needed. Some patients may require periodic ECG or Holter monitoring to determine the likelihood and rate of progression of the conduction disorder. A permanent pacemaker is indicated in cases of infranodal block, suggested by AV block at intra- or infra-His levels, prolonged HV conduction (HV interval > 100 milliseconds), and pacing-induced infrahisian AV block.

6. Cardiology evaluation

Cardiology evaluation is usually not necessary, indicated only in the rather rare case of infranodal block requiring permanent pacing.

B. Type II Second-degree AV Block

1. General principles

Mobitz II second-degree AV block is characterized by abrupt and periodic failure of an atrial beat to be conducted to the ventricle, without Wenckebach periodicity. It is commonly associated with significant conduction system disease, with risk of more advanced block. Its diagnosis usually implies an indication for permanent pacing.

2. Mechanisms

This type of block is very rare in healthy individuals. It most often occurs in the His-Purkinje system (20% in the bundle of His), and it often progresses to third-degree block. Common causes are:

- **Acute MI/ischemia:** In case of an inferior infarct the block is usually nodal and reversible, whereas anterior infarction may lead to block due to ischemia/infarction of bundle branches.
- **Degenerative changes in the AVN or bundle branches:** These account for > 80% of cases of AV block; classic progressive conditions are Lev's (idiopathic fibrosis of the cardiac skeleton) and Lenegre's disease (idiopathic fibrosis of the distal conduction system).
- **Infiltrative myocardial diseases:** Sarcoidosis, myxedema, hemochromatosis.
- **Progressive calcification of the aortic or mitral annular rings.**
- **Infections of the myocardium:** Lyme disease, for example.
- **Systemic diseases**: Ankylosing spondylitis and Reiter syndrome.
- **Procedures:** Surgical (e.g., aortic valve replacement, closure of septal defects) or other invasive procedures (e.g., ablation in patients with supraventricular tachycardia [SVT], alcohol septal ablation in patients with hypertrophic cardiomyopathy [HCM]).
- **Drug influence:** Amiodarone, quinidine, procainamide, etc.

3. Clinical history

Some patients are asymptomatic; others may feel irregularities of the heartbeat. Presyncope and syncope (Stokes-Adams type) suggest episodes of higher-degree AV block.

4. Diagnosis

a. Electrocardiogram (Figure 4.2): Diagnosis by ECG includes conducted P waves, with stable PR, alternating with a blocked P; the PR of the first conducted beat after the blocked P is similar as in the previous conducted beats; the pause after the blocked P wave equals the sum of the 2 preceding beats; wide QRS complexes,

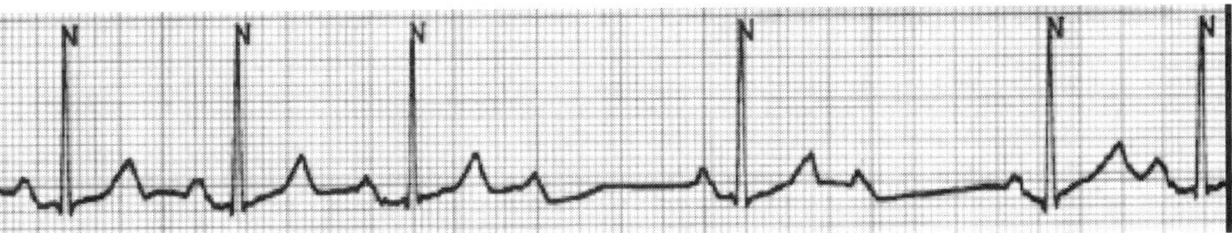

Figure 4.2. Mobitz II second-degree AV block.

indicative of infrahisian block—rarely, the block is isolated to the His bundle and, therefore, the QRS is narrow.

b. Laboratory testing: As in type I second-degree AV block.

c. EP Study: An EPS, rarely necessary, may be of use if the site of the block is not evident from analysis of the surface ECG—for example, asymptomatic type II block with narrow QRS. The test will reveal the site of the block, almost always infranodal, with marked prolongation of the HV interval (or split H potential) and a regular blocked AH (not followed by ventricular activation), without previous HV prolongation.

Identifying episodes of transient AV block with sudden pauses and/or low HR, causing syncopal episodes, may require Holter monitoring, periodic ECG recording, external event recorders, or, rarely, monitoring with implantable loop recorders.

5. Therapy

Affected patients require hospitalization, with continuous rhythm monitoring. Correctible causes should be identified (e.g., acute ischemia, drugs that slow AV conduction) and managed accordingly. Transcutaneous or transvenous temporary pacing should be used in acutely symptomatic patients while awaiting pacemaker implantation (Table 4.1).

In cases of acute inferior infarction, almost all types of AV block are located at the AVN level and there is usually no indication for permanent pacing; if symptomatic, IV atropine or temporary pacing can be used. Recommendations for permanent pacing after the acute phase of an infarction are found in Table 4.2.

6. Cardiology evaluation

Cardiac evaluation is recommended as these patients are candidates for permanent pacing.

Table 4.1. Recommendations for Permanent Pacing in Acquired AV Block in Adults

Class I

- Third-degree and advanced second-degree AV block at any anatomic level associated with symptomatic bradycardia or ventricular arrhythmia presumed to be due to AV block
- Third-degree and advanced second-degree AV block at any anatomic level associated with arrhythmias or other conditions that require drugs that result in symptomatic bradycardia
- Third-degree and advanced second-degree AV block at any anatomic level in awake, symptom-free patients in sinus rhythm, with documented asystole ≥ 3 s, any escape rate < 40 bpm, or with an escape rhythm that is below the AVN
- Third-degree and advanced second-degree AV block at any anatomic level in awake, symptom-free patients with AF and bradycardia with 1 or more pauses ≥ 5 s
- Third-degree and advanced second-degree AV block at any anatomic level after catheter ablation of AVN
- Third-degree and advanced second-degree AV block at any anatomic level associated with postoperative AV block that is not expected to resolve after cardiac surgery
- Third-degree and advanced second-degree AV block at any anatomic level associated with neuromuscular diseases with AV block (e.g., myotonic dystrophy, Erb's, peroneal muscular atrophy, Kearns-Sayre), with or without symptoms
- Second-degree AV block with associated symptomatic bradycardia regardless of type or site of block
- Asymptomatic persistent third-degree AV block at any anatomic site with average awake ventricular rate ≥ 40 bpm if cardiomegaly or LV dysfunction is present or if the site of block is below the AVN
- Second- or third-degree AV block during exercise in the absence of myocardial ischemia

Class IIa

- Persistent third-degree AV block with escape rate > 40 bpm in asymptomatic adult patients without cardiomegaly
- Asymptomatic second-degree AV block at intra- or infra-His levels found at EPS
- First- or second-degree AV block with symptoms similar to those of pacemaker syndrome or hemodynamic compromise

Table 4.1. *(Continued)*

- Asymptomatic type II second-degree AV block with a narrow QRS (presence of a wide QRS, including isolated RBBB, makes it a class I recommendation)

Class IIb

- Neuromuscular disease (e.g., as above) with any degree of AV block (including first-degree AV block), with or without symptoms
- AV block in the setting of drug use and/or drug toxicity when the block is expected to recur even after the drug is withdrawn

Class III

- Asymptomatic first-degree AV block
- Asymptomatic type I second-degree AV block at the supra-His level or not known to be intra- or infrahisian
- AV block expected to resolve and/or unlikely to recur (e.g., drug toxicity, Lyme disease, hypoxia in obstructive sleep apnea in the absence of symptoms)

IV. 2:1 AV BLOCK

A 2:1 AV block is the presence of 2 sinus P waves for every 1 QRS complex (Figure 4.3). It may be due to AV nodal or infranodal disease; clinically, proper location of the site of block is always difficult, and it requires intracardiac recordings for a positive identification (occasionally IV atropine helps in the diagnosis, inducing a 3:2 AV conduction). By itself, a 2:1 AV block cannot be classified as a type I or II mechanism because only one PR interval is available for analysis before the block (Table 4.3). The co-existence of a prolonged QRS interval suggests however that the location of the block is infrahisian.

Both a 2:1 AV block and a block involving 2 or more consecutive sinus P waves are sometimes referred to as advanced or high-degree AV block. Their natural history is similar to that of type II second-degree AV block and should be managed in a similar fashion.

V. THIRD-DEGREE AV BLOCK

A. General Principles

Independent atrial and ventricular complexes characterize third-degree AV block, with atrial rate usually exceeding the ventricular rate (30–40 bpm), and cardiac

Table 4.2. Recommendations for Permanent Pacing After the Acute Phase of Myocardial Infarction

Class I

- Persistent second-degree AV block in the His-Purkinje system with alternating BBB or third-degree AV block within or below the His-Purkinje system after ST-elevation myocardial infarction (STEMI)
- Transient advanced (second- or third-degree) infranodal AV block and associated BBB (if site of block is uncertain EPS may be necessary)
- Persistent and symptomatic second- or third-degree AV block

Class IIb

- Persistent second- or third-degree AV block at the AVN level, even if asymptomatic

Class III

- Transient AV block in the absence of intraventricular conduction defects
- Transient AV block in the presence of isolated left anterior fascicular block
- New BBB or fascicular block in the absence of AV block
- Persistent asymptomatic first-degree AV block in the presence of bundle branch or fascicular block

rhythm depending on a subsidiary pacemaker. Its prevalence in the general population is 0.02% to 0.04%. Third-degree AV block may be an underlying condition in patients who present with syncope/sudden cardiac death (SCD).

B. Mechanisms

The etiology of third-degree AV block is similar to that of type II second-degree AV block. Congenital third-degree AV block is rare (1 case per 20,000 births), and the site of block is typically nodal: in the absence of major SHD, neonatal lupus is associated in 60% to 90% of these cases. Its mortality approaches 20% and most surviving children require a permanent pacemaker. Most cases of complete heart block in adults are acquired and due to degeneration of the His-Purkinje system.

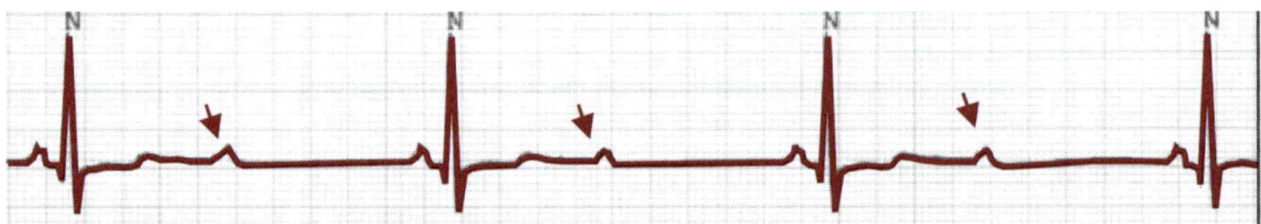

Figure 4.3. 2:1 AV block. Arrows indicate the blocked P waves.

Table 4.3. ECG Features Indicating Origin of Block

Nodal Origin	Infranodal Origin
Periods of type I second-degree AV block	Concomitant type II second-degree AV block
PP prolongation preceding the AV block	No evident PP prolongation preceding the AV block
Narrow QRS complexes	Wide QRS complexes

C. Clinical History

Some patients are asymptomatic, especially when the block is at the AVN level. Bilateral bundle branch disease is usually very symptomatic, with fatigue, dizziness, light-headedness, presyncope, and syncope. Complications resulting from third-degree AV block are SCD due to asystole or secondary ventricular tachyarrhythmias, cardiovascular collapse with syncope, aggravation of myocardial ischemia, HF, exacerbation of renal disease, and head and musculoskeletal injuries during syncopal episodes. In cases of congenital third-degree AV block fetal bradycardia is usually the presenting sign; it may be intermittent when first detected but usually becomes persistent in later childhood.

D. Physical Examination

A physical examination will typically find severe bradycardia, with slow pulse; intermittent cannon *a* waves; variable S_1 intensity; wide BP; and secondary signs, such as dyspnea, confusion, and pulmonary edema.

E. Diagnosis

1. Electrocardiogram (Figure 4.4)

Diagnosis by ECG will find complete independence between atrial and ventricular contractions—no P wave originates a QRS complex; atrial rate that exceeds the ventricular rate; and very regular R-R interval (if the block is not complete, e.g., high-degree AV block, there could be some irregularity of the R-R). Wide QRS complexes indicate an infranodal origin, while narrow QRS (< 120 milliseconds) indicate block at the nodal junction.

2. Laboratory testing

As in type I second-degree AV block.

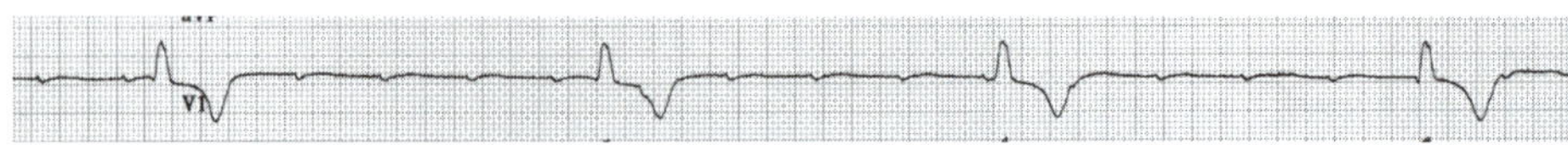

Figure 4.4. Complete AV block.

3. Imaging

An echocardiogram is indicated if SHD is suspected.

4. EP Study

Not necessary.

F. Therapy

Therapy is similar to that of type II second-degree AV block. Once the patient is stable, a permanent pacemaker implantation is usually indicated (see Tables 4.1 and 4.2).

G. Cardiology Evaluation

Recommended, as most patients end up requiring permanent pacing.

VI. AV DISSOCIATION

General Principles

AV dissociation occurs when there is no synchronism in AV conduction; it is occasionally benign and its presence does not necessarily indicate an underlying AV block. It is never a primary disorder, and it is always necessary to determine its origin.

Complete AV dissociation can mimic AV block, but the fact that none of the P waves conduct has more to do with timing of the P waves in relation to the QRS complex than with the presence of block. In its incomplete presentation, some of the P waves conduct and capture the ventricles (interference AV dissociation). When independent atrial and ventricular rhythms are present at nearly identical rates, it is called isorrhythmic dissociation (a junctional rhythm should be excluded).

Mechanisms

The pathophysiology of AV dissociation also helps in its classification:

- **AV dissociation by default:** Due to slowing of the sinus rate, for example during sinus arrhythmia or sinus bradycardia.
- **AV dissociation by usurpation:** Due to increased discharge rate of a usually slower subsidiary pacemaker—for example, nonparoxysmal AV junc-

tional tachycardia, accelerated idioventricular rhythm, junctional rhythm/ VT without retrograde atrial capture.

- **AV block.** A complete AV block always leads to AV dissociation.

C. Clinical History

AV dissociation can be asymptomatic; when symptoms are present, they are usually limited to exertional dyspnea, light-headedness, palpitations, fatigue, and general discomfort.

D. Physical Examination

A physical examination will reveal the following: pulse volume is variable, with fast or slow rates depending on the underlying cause; low BP in VT; intermittent cannon *a* waves; variable S_1 intensity; beat-to-beat variation in systolic murmurs.

E. Diagnosis

1. Electrocardiogram (Figure 4.5)

Diagnosis by ECG will include ventricular rate the same or faster than the atrial rate. No retrograde (VA) conduction occurs.

Determining whether there is AV conduction or not may be difficult: maneuvers that modify the atrial and ventricular rates may help.

2. Laboratory testing

Digoxin level should be measured in all patients on digitalis therapy.

3. Imaging

An echocardiogram is indicated if SHD is suspected.

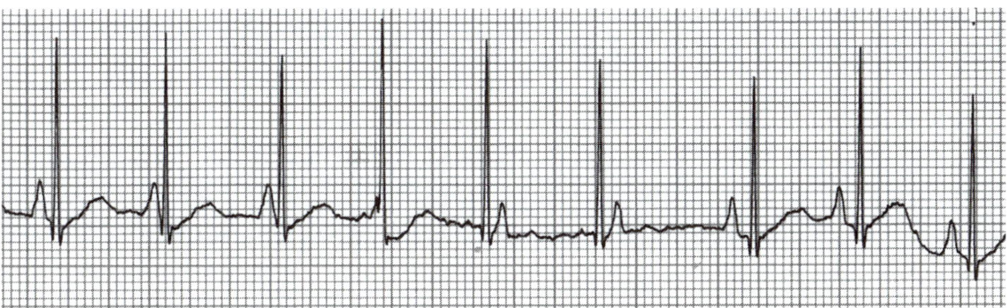

Figure 4.5. AV dissociation by usurpation. Acceleration of the AVN overtakes the SAN from the fourth to the sixth beats.

4. EP Study

Usually not required.

F. Therapy

Treatment depends on the underlying condition and its severity, as most episodes resolve with resolution of the causative factor. Patients who are hemodynamically unstable—for example, patients with VT—should be treated accordingly. If necessary, acceleration of the HR can be attained using atropine or isoproterenol. Digoxin immune Fab fragments are indicated in treatment of digoxin toxicity.

Ascertaining if AV conduction is intact is important in patients with AV dissociation due to an accelerated junctional rhythm following cardiac surgery: some patients may have a complete AV block with an accelerated focus distal to the level of block (when the accelerated focus slows down the heart block then becomes evident).

G. Cardiology Evaluation

Cardiac consult is needed in cases of unexplained or uncorrected persistent symptomatic AV dissociation due to an escape rhythm or VT.

VII. FASCICULAR BLOCKS

A. General Principles

Fascicles of the AV conduction system can be affected separately or in conjunction.

Bifascicular block refers to a conduction block below the AVN in which the right bundle branch and one of the fascicles of the left bundle branch are involved. The term *trifascicular block* is usually reserved for the combination of bifascicular block and prolongation of the PR interval. Alternating BBB (e.g., RBBB and LBBB on successive ECGs) is diagnostic of severe diffuse distal conduction disease and is an indication for permanent pacing; the other possible setting, the combination of RBBB and block of both fascicles of the left bundle documented at the same time, will usually manifest as third-degree AV block.

In chronic bifascicular and trifascicular blocks the annual progression to complete heart block is uncommon but the actual rate differs significantly among published studies.

Diagnosis

1. Electrocardiogram

a. Left anterior fascicular block: Left-axis deviation with frontal QRS axis between –45° and –90°; Q wave in aVL; rS morphology in inferior leads; QRS duration < 120 milliseconds.

b. Left posterior fascicular block: Right-axis deviation with frontal QRS axis between +90° and +180°; rS in leads I and aVL; qR morphology in inferior leads (Q wave ≤ 40 milliseconds); QRS duration <120 milliseconds. For the diagnosis of left posterior fascicular block, other causes of right-axis deviation must be first excluded.

2. EP Study

An EPS is considered only in patients who have otherwise unexplained presyncope/syncope; in such patients, the following findings are possible indications for pacemaker implantation: AV block at intra- or infra-His levels as opposed to within the AVN; markedly prolonged HV interval, > 100 milliseconds; pacing-induced infrahisian block.

C. Therapy

Isolated fascicular blocks are asymptomatic. In selected patients, treatment may include correction of reversible causes like ischemia, avoiding drugs that impair AV conduction, and permanent pacing when indicated (Table 4.4).

Table 4.4. Recommendations for Permanent Pacing in Chronic Bifascicular Block

Class I

- Advanced second-degree AV block or intermittent third-degree AV block
- Type II second-degree AV block
- Alternating BBB

Class IIa

- Syncope not demonstrated to be due to AV block when other likely causes have been excluded, specifically VT
- Incidental finding at EPS of prolonged HV interval ≥ 100 ms in asymptomatic patients
- Incidental finding at EPS of pacing-induced infra-His block that is not physiologic

Class IIb

- Neuromuscular diseases (e.g., myotonic dystrophy, Erb's, peroneal muscular atrophy, Kearns-Sayre) with any degree of fascicular block, with or without symptoms

Class III

- Fascicular block without AV block or symptoms
- Fascicular block with first-degree AV block without symptoms

D. Cardiology Evaluation

Cardiac consult is suggested for patients meeting criteria for permanent pacing.

SUGGESTED READINGS

Epstein AE, Dimarco JP, Ellenbogen KA, et al. ACC/AHA/HRS 2008 guidelines for device-based therapy of cardiac rhythm. *J Am Coll Cardiol.* 2008;51:2085-2105.

PART 3

Syncope, Sudden Cardiac Death, and Clinical Arrhythmias

CHAPTER 5

Syncope

Miguel A. Barrero Garcia and Laurent Macle

I. GENERAL PRINCIPLES

Syncope is a transient loss of consciousness, usually associated with loss of postural tone, due to transient global cerebral hypoperfusion characterized by rapid onset, short duration, and spontaneous complete recovery.

According to the Framingham study, the incidence of new syncope is 6.2 per 1000 persons per year, being almost double in patients with cardiovascular disease (when compared to those without it). Among syncope patients recurrence is common (30%). Neurally mediated mechanisms account for 35% to 58% of all causes of syncope while cardiac causes are found in about 20%; in 18% of patients the syncope is unexplained. The prognostic significance of syncope depends on its cause, the nature and severity of underlying structural heart disease (SHD), and the type of treatment initiated. Mortality is higher in patients with left ventricle (LV) dysfunction.

Even in cases with a benign etiology, syncope could be associated with significant traumatic injury, particularly in the elderly.

II. MECHANISMS

Syncope results from the transient failure of protective mechanisms that maintain blood pressure (BP) and guarantee brain perfusion. Its causes can be broadly classified as cardiovascular and noncardiovascular (Figure 5.1). Among cardiovascular causes, electrical disorders are the most important. Among noncardiovascular etiologies, neurally mediated syncope is the commonest: its most accepted mechanism is that in patients who are predisposed to have increased peripheral venous pooling, a sudden drop in preload leads to a hypercontractile state, which in turn activates mechanoreceptors that provoke a false central nervous system perception of hypertension, resulting in bradycardia, vasodilatation, or both. Similar mechanoreceptors

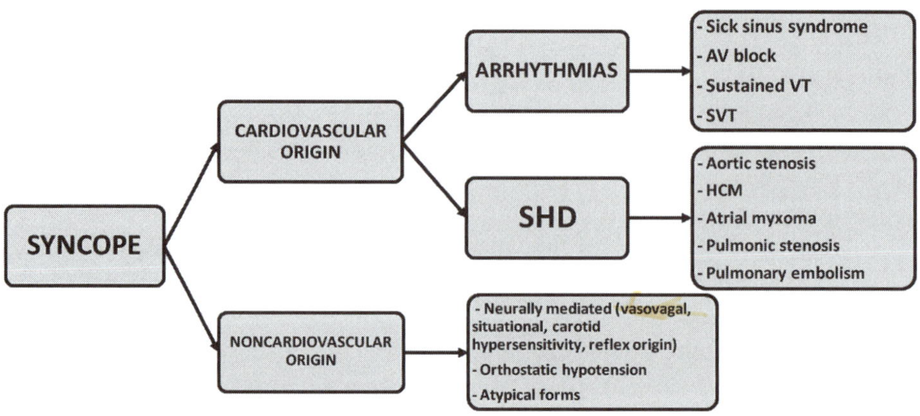

Figure 5.1. Causes of syncope.

are also present in other parts of the body such as the bladder, rectum, esophagus, and lungs, thus explaining reflex syncope due to micturition, defecation, deglutition, and cough, respectively.

In some cases a specific etiology is not determined despite full evaluation including electrophysiological study (EPS) (syncope of unknown origin). Syncope-like conditions are usually psychogenic, like hyperventilation.

III. CLINICAL HISTORY

History taking is extremely important, as in close to 50% of cases it helps determine the cause of syncope. The report of a reliable eyewitness may be very helpful as affected patients may not be able to remember or may overlook important details. A family history of recurrent syncope/sudden cardiac death (SCD) should always be elucidated. The presence of cardiac conditions is a strong predictor of cardiac causes. Recurrent syncope over many years is usually benign; unexplained falls, particularly in elderly persons, may also be due to syncope. Trauma associated to syncope is common.

Patients may describe a syncopal episode in many ways, including blackout, dizzy spell, and seizure. If no loss of consciousness can be confirmed, presyncope is the appropriate term to use. Other associated symptoms depend on the etiology, and in many cases they help confirm the clinical diagnosis. Vasovagal syncope is usually preceded by progressive light-headedness, diaphoresis, or nausea (often triggered by fear, repulsion, or anxiety) and the loss of consciousness is followed by fatigue and tiredness for several hours. Syncope of cardiac origin is usually not related to changes of position, there are no prodromes, and the patients tend to recover quite rapidly; preceding palpitations should make a tachyarrhythmia a strong suspect. Associated chest pain or shortness of breath should be thoroughly investigated. A history of focal neurological symptoms or sphincter incontinence should also be determined.

The clinical differential diagnosis of syncope is extremely important:

- **Vertigo:** There is no real passing out but a sensation of movement of the patient and/or the surroundings. Its etiology is usually neurologic or otolaryngologic. Vertigo may be reproduced with the Parinet maneuver.
- **Seizure:** Most seizures are characterized by sudden loss of postural tone followed up by tonic-clonic convulsions. Incontinence and confusion postevent (postictal state) are very typical. Some rare types of seizures (e.g., partial complex) without significant motor activity, and syncope with tonic posturing due to brainstem hypoxia, are more difficult to differentiate clinically.
- **Transient ischemic attack:** A focal neurological defect is frequently found. When it is of vertebrobasilar origin it may rarely cause loss of consciousness but other vertebrobasilar symptoms are usually present (e.g., dysarthria, dysphagia).

- **Narcolepsy:** Sudden episodes of sleep during the day, with no real unconsciousness.
- **Pseudosyncope:** A faked episode, usually intent on secondary gain. A psychiatric disorder should be excluded.
- **Cataplexy:** A rare condition characterized by transient loss of postural tone without loss of consciousness.

IV. PHYSICAL EXAMINATION

Normal findings are not necessarily an indication of benignity as many life-threatening tachyarrhythmias occur in otherwise healthy people. Similarly, a trauma speaks more of a real episode of syncope than to the etiology of it. When found, cardiovascular and focal neurologic abnormalities help establish the diagnosis. Postural hypotension should always be excluded (at least a 20 mm Hg decrease in systolic BP, a 10 mm Hg decrease in diastolic BP, or symptoms, associated with increase in heart rate [HR] in the absence of sick sinus syndrome [SSS] or negative chronotropic drugs). A carotid sinus massage (CSM) should be performed in all patients without contraindications (see Chapter 13).

V. DIAGNOSIS

A. Electrocardiogram

A rest ECG could reveal abnormalities such as varied degrees of atrioventricular (AV) block, long QT interval, ventricular preexcitation, signs of ischemia/infarction, Brugada sign, or signs of arrhythmogenic right ventricle cardiomyopathy/dysplasia (ARVC/D). In patients with neurally mediated syncope, the ECG is often normal.

B. Laboratory Testing

Electrolytes, digoxinemia, cardiac enzymes, and thyroid function tests may be needed if there is clinical suspicion of their respective underlying disorders.

C. Imaging

Transthoracic echocardiography is routinely performed in most patients because of the need to exclude SHD (LV dysfunction, outflow tract obstructions, etc.).

D. Other

Other tests are performed depending on the suspected etiology:

1. Cardiac origin

a. EST: Of utility in patients with suspicion of exercise-induced arrhythmias or if syncope is related to physical exercise (care should be taken to first exclude severe aortic stenosis and hypertophic cardiomyopathy [HCM]).

b. Coronariography: It should be carried out in patients with suspected myocardial ischemia/infarction and to rule out ischemia-driven arrhythmias.

c. Ambulatory ECG monitoring: Its utility depends on the frequency of symptoms. Holter monitoring and external loop recorders help diagnose rhythm disorders as the cause of frequent unexplained syncope, but they have a very low yield in patients with infrequent episodes; in these cases an implantable loop recorder may be indicated (yield at 5 months, 65%–88%).

d. EP Study: May be beneficial in selected patients, mostly those with syncope of unknown cause with impaired LV function or SHD, or when a bradyarrhythmia or tachyarrhythmia is suspected and noninvasive studies have been inconclusive.

e. Measurement of plasma BNP: May be useful in distinguishing cardiac from noncardiac causes of syncope.

Occasionally some other forms of testing are needed—for example, signal-averaged ECG (SAECG) or T-wave alternans (TWA).

2. Suspected Neurally Mediated Syncope

CSM and the tilt table test (TTT) are appropriate diagnostic approaches. If vasovagal syncope is strongly suspected clinically and no other risk factors are present,

Table 5.1. Markers of High Risk in Syncope

Underlying SHD, especially LV dysfunction
Exertional syncope
Family history of sudden death
Significant associated trauma
Nonsustained ventricular tachycardia (NSVT)
Bifascicular block or QRS duration ≥ 120 ms
Inadequate sinus bradycardia or SA block
Ventricular preexcitation
Long or short QT interval
Brugada pattern, or signs of ARVC/D in the ECG

confirmation with a TTT is not necessary in most patients as the test could be negative despite a strong vagal component (see Chapter 13 for more information).

3. Neurologic Origin

Neurologic testing (electroencephalogram, brain CT/MRI, carotid Doppler ultrasonography) is often performed in patients with unexplained syncope, but their utility is quite limited unless the patient has syncope with associated focal neurological deficits or if there is witnessed seizure activity.

A common practice in the initial evaluation of unexplained syncope is to perform a CSM, followed by TTT and echocardiogram. The physician should always be on the lookout for high-risk features that demand intensive evaluation or hospitalization (Table 5.1). Until the diagnosis is established, and successful treatment put in place, driving should be prohibited.

VI. THERAPY

Management of patients with syncope depends on the underlying etiology and is directed at preventing recurrence and/or death.

A. Neurally Mediated Syncope

Conditions that trigger episodes of vasovagal syncope should be avoided. Rare episodes with a specific trigger (e.g., drawing blood) generally do not require specific therapy. Nonpharmacological therapy is often adequate in more symptomatic patients, including avoidance of triggering situations, preventing dehydration, liberalizing salt intake, and wearing support stockings to reduce venous pooling. Discontinuation of offending medications (e.g., nitrates) that are not of obligatory use may also be helpful. Orthostatic training programs can also be tried. Pharmacological therapies have been employed in managing very symptomatic cases (Table 5.2), but, so far, there is no firm evidence for drug efficacy.

Permanent pacing may be considered in patients with cardioinhibitory neurocardiogenic syncope and carotid sinus hypersensitivity, in whom other medical therapy is ineffective or not indicated.

B. Orthostatic Hypotension

Orthostatic hypotension is occasionally diagnosed in patients with presyncope/syncope on changes of position (e.g., from sitting down to standing up). In young patients it may reflect dysautonomia; in elderly patients, it may be the result of conditions such as diabetes or medications. Nonpharmacologic measures include avoidance of volume depletion and medications such as β-blockers and antidepressants. Helpful physical measures are leg exercising (i.e., leg tensing) before arising slowly in

Table 5.2. Drugs for Control of Recurrent Vagal Syncope

Drugs	Dose
Hydrofludrocortisone	0.1 mg PO bid
Metoprolol	50–100 mg PO bid
Disopyramide	150–200 mg PO bid
Theophylline	200 mg PO bid
Ritalin	60 mg PO qid, or 5–15 mg PO tid
Midodrine	2.5–10.0 mg PO tid
Sertraline	25–100 mg PO qid
Nefazodone	100–150 mg PO tid

stages from supine to seating to standing, use of compression stockings, and head-up tilt sleeping (> 10°). Drug therapy may include midodrine, fludrocortisone, caffeine, and fluoxetine. Patients on fludrocortisone and/or liberal salt intake must be monitored carefully for the development of edema or hypertension.

C. Structural Heart Disease

Treatment of patients with syncope due to SHD is directed at the underlying cardiac pathology. In aortic stenosis, syncope usually indicates a valve replacement is needed. Patients with HCM may benefit from treatment with β-blockers, Ca^{2+} channel blockers (CCBs), myomectomy, or implantable cardioverter defibrillator (ICD).

D. Bradyarrhythmias

A permanent pacemaker is indicated when there is a sinus node dysfunction (SND) or AV block documented to cause syncope. Occasionally, a pacemaker is implanted empirically in older patients with recurrent syncope and no documentation of SND/AV conduction disorders, and/or if the EPS strongly suggests a conduction abnormality as the cause for syncope.

E. Ventricular Tachyarrhythmias

Patients with SHD or conditions such as Brugada syndrome (BrS), long QT syndrome (LQTS), cathecolaminergic polymorphic ventricular tachycardia (CPVT), ARVC/D, and unexplained syncope who have documentation of an episode of sustained ventricular tachycardia (VT), ventricular fibrillation (VF), or inducible

VT/VF during EPS are candidates for ICD implantation. For those in whom ICDs are not appropriate or desired, treatment with amiodarone, sotalol, or dofetilide is also an option (LQTS often responds to β-blockers).

Catheter ablation is effective, especially in patients with idiopathic monomorphic VT.

F. Supraventricular Tachycardia

In patients with supraventricular tachycardia (SVT) causing syncope or patients with Wolff-Parkinson-White (WPW) syndrome and syncope, catheter ablation of the arrhythmia circuit is preferred over treatment with antiarrhythmic drugs (AADs).

G. Cardiomyopathy and Unexplained Syncope

Patients with cardiomyopathy and LV dysfunction having syncope, regardless of cause, have a high risk of sudden death. Additionally, syncope in this setting is most likely of arrhythmic nature. They should all be evaluated thoroughly (including EPS if necessary), medical therapy should be optimized, and, if eligibility criteria are met (see Chapter 12), ICD therapy pursued for the purpose of prevention of SCD.

VII. CARDIOLOGY EVALUATION

Patients with recurrent syncope or syncope in the setting of significant SHD should be referred for prompt cardiac consultation; same applies when an arrhythmic cause is suspected.

SUGGESTED READINGS

Moya A, Sutton R, Ammirati F, et al. Guidelines for the diagnosis and management of syncope (version 2009). *Eur Heart J.* 2009;30(21):2631-2671.

Strickberger SA, Benson DW, Biaggioni I, et al. AHA/ACCF scientific statement on the evaluation of syncope. *J Am Coll Cardiol.* 2006;47:473-484.

CHAPTER 6

Sudden Cardiac Death

Miguel A. Barrero Garcia and Mario Talajic

I. GENERAL PRINCIPLES

Sudden cardiac death (SCD) is a natural and unexpected death due to cardiac causes that occurs within 1 hour of the onset of symptoms. It is a leading cause of mortality, with an estimated annual incidence in the United States of 200,000 to 450,000 cases. Its incidence increases as patients age, with a peak between 45 and 75 years of age and a male/female ratio of 3:1. A higher incidence of SCD occurs among certain high-risk groups (Figure 6.1), but these groups represent only a small percentage of total events occurring in the general population.

Approximately 60% of SCDs take place outside a hospital setting, a fact that underlies the importance of adequate prehospital recognition and management protocols.

II. MECHANISMS

A. Arrhythmic Mechanism

The arrhythmic mechanism of SCD is different from that of sustained ventricular tachycardia (VT) due to reentry. Sustained monomorphic VT can degenerate into ventricular fibrillation (VF), but VF or polymorphic VT are usually the arrhythmias

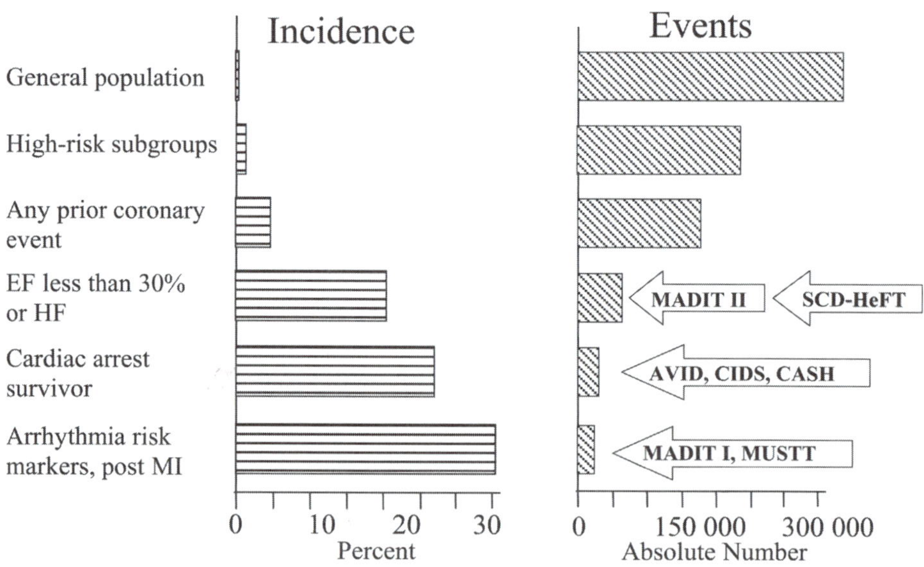

Figure 6.1. Absolute number of events and event rates of SCD in the general population and in specific subpopulations over 1 y (the arrows indicate trials carried out in the respective groups of patients). (Reprinted with Permission from Zipes DP, Camm AJ, Borggrefe M, et al. ACC/AHA/ESC 2006 guidelines for management of patients with ventricular arrhythmias and the prevention of sudden cardiac death. *Circulation* 2006;114:e385-e484. © 2006 American Heart Association, Inc.)

present at the onset of an episode of cardiac arrest, likely the result of an interaction between an abnormal substrate and transient and functional factors (e.g., worsened ischemia, acidosis, hypoxemia, wall tension, drug use, metabolic disturbances). Approximately 20% to 30% of patients with a documented SCD present with a bradyarrhythmia, asystole, or pulseless electrical activity.

Most cases of arrhythmic SCD occur in patients with structural heart disease (SHD). The majority of episodes occur in patients with symptomatic heart failure (HF) and associated left ventricle (LV) dysfunction; these patients have a 20% to 25% increased risk of premature death in the first 2.5 years after diagnosis, and it is estimated that 50% of these deaths are sudden and due to VT/VF. Malignant ventricular arrhythmias may also complicate the initial hours after a myocardial infarction (MI) or occur during episodes of reversible myocardial ischemia.

Less commonly, SCD occurs in patients with inherited arrhythmia syndromes; these syndromes, which have a genetic basis, may be associated with a structurally normal heart or with a genetically determined cardiomyopathy. Atrial fibrillation (AF) in the setting of the WPW syndrome may be associated with very rapid ventricular conduction, which may degenerate into VT/VF and subsequent sudden death.

B. Nonarrhythmic Causes

Among nonarrhythmic causes of SCD, pulmonary embolism and aortic dissection are the commonest. Pulmonary embolism is a frequent cause of sudden death in people with risk factors such as history of deep vein thrombosis, malignancy, hypercoagulable states, and recent mechanical trauma like hip or knee surgery. Predisposing factors for aortic dissection include Marfan and Ehlers-Danlos syndromes, and aortic cystic medial necrosis.

III. CLINICAL HISTORY

Syncope, sometimes recurrent, is present in some patients; when there is a family history of SCD it suggests an inherited arrhythmia syndrome. A history of previous successful resuscitation from a cardiac arrest, a history of VT, and the presence of LV dysfunction are the most common risk factors for SCD.

Though obviously unexpected, patients with SCD may have some preceding complaints (chest pain, fatigue, palpitations, dizzy spells), often nonspecific.

IV. PHYSICAL EXAMINATION

The physical examination may reveal evidence of the underlying cardiovascular condition, or it may be entirely normal. In survivors, the use of a cardiac arrest score helps predict the clinical evolution: patients with higher scores have a good chance of neurologic recovery and survival to discharge, while severe anoxic encephalopathy (30%–80% of cases) in patients with low scores indicates a poor survival rate and consequently management has been usually supportive and noninvasive (though the use of therapeutic cooling in selected patients has been found to reduce mortality by 40%).

V. DIAGNOSIS

A. Risk Stratification

Several markers have been used to stratify the risk for SCD (Table 6.1); individually, they all lack high sensitivity and specificity. So far there is no single best measurement to reliably identify all patients at risk for SCD. Some tests, like ECG and echocardiography, are indicated in most patients; others have specific indications.

B. Laboratory Testing

1. Cardiac enzymes

Elevations may indicate an acute coronary syndrome. Even though there is an increased risk for arrhythmia in the peri-infarct period (incidence of VT/VF of 10% in patients with ST-elevation myocardial infarction [STEMI] and 2% for non ST-elevation myocardial infarction [NSTEMI]), it usually does not imply long-term

Table 6.1. Markers of Risk for SCD

Parameter	Marker of Risk	Utility
Medical history	Family history of sudden death, syncope, known cardiopathies	All patients
ECG	NSVT, necrosis, short or long QT, Brugada sign, ventricular preexcitation, short R-R interval during AF	CAD, LQTS, BrS, WPW syndrome, AF
TWA	Positive TWA	Myocardial infarction
SAECG	Positive late potentials	ARVC/D
EPS	Short antegrade accessory pathway RP, inducible sustained VT	WPW, LV dysfunction, BBR, Fallot tetralogy, HCM
Echocardiogram	Low EF, asymmetric LVH	DCM, HCM
Genetic testing	Varied markers	LQTS, BrS, HCM, ARVC/D

increased cardiovascular risk as most arrhythmias are probably related to ischemia or reperfusion.

2. Electrolytes

Risk for arrhythmia and sudden death increases in severe metabolic acidosis, hypokalemia, hyperkalemia, hypocalcemia, and hypomagnesemia.

3. Drugs

AADs and other drugs may lead to an increased risk for ventricular arrhythmia, especially AADs due to their potential proarrhythmic effect.

4. BNP measurement

Measurement of BNP (a peptide secreted mainly from the ventricles in response to ventricular strain) has a predictive value, especially in post-MI patients and in patients with HF. It seems to provide prognostic information on the risk of SCD, independent of clinical information and LV ejection fraction (EF).

C. Imaging

1. Echocardiography

The resting EF is the most important known noninvasive predictor of SCD and other cardiac events in patients with MI. The presence of segmental abnormalities in patients with a history of syncope/cardiac arrest should raise the suspicion of a reentrant VT as the causative factor. Septal thickness > 30 mm increases the risk of sudden death in hypertrophic cardiomyopathy (HCM).

2. Nuclear imaging

Nuclear imaging serves to complement information provided by echocardiography. Myocardial perfusion imaging is very sensitive for detecting the presence, extent, and location of myocardial ischemia. Pharmacological stress (dipyridamole or adenosine) has been found better than submaximal exercise stress testing (EST) and coronariography in predicting cardiac death and other cardiac events. In patients with suspected ARVC/D, MRI provides more information that regular imaging techniques.

3. Coronary angiography

Coronary angiography is recommended in all patients who survive an episode of cardiac arrest to assess the degree of severity of coronary artery disease (CAD), possible coronary anomalies, and the state of ventricular function. In candidate patients revascularization is the best treatment for the underlying substrate of VT/VF: ischemic myocardium.

VI. THERAPY

A. Medical Management

Because up to 40% of cases of cardiac arrest are unwitnessed, survival in many cases depends on adequate prehospital management protocols, centralized on basic life support/ advanced cardiac life support (BLS/ACLS) by appropriately trained individuals, quick response from emergency services, and rapid transfer to a hospital. The use of automatic external defibrillators in public places (airports, supermarkets, etc.) has improved the outcome of witnessed cardiac arrest.

Progress in medical therapy has reduced significantly the risk of SCD in selected groups of patients. The use of β-blockers and angiotensin-converting enzyme inhibitors (ACEIs) decreases the risk of sudden death and overall mortality. AADs are useful in controlling symptomatic arrhythmias but they have not been proven conclusively to be superior to placebo in preventing sudden death.

B. Nonpharmacological Therapy

1. Catheter ablation

Ablation is of utility in patients with conditions such as accessory pathways, bundle branch reentry (BBR) VT, idiopathic LV VT. Unfortunately, many cases of arrhythmic cardiac arrest are not amenable or do not respond well to ablation. In most cases VT ablation is palliative and does not eliminate the need for other therapy such as the implantation of an implantable cardioverter defibrillator (ICD).

2. ICD

In terms of secondary prevention, ICDs are superior to AADs. Their effect on improving survival is quite significant, resulting in a 28% reduction in the risk of death (due chiefly to a 50% reduction in arrhythmic mortality).

In primary prevention, ICD implants have been proven to confer a significant benefit in patients with low EF (< 30%–35%) and are an accepted standard therapy. The annual SCD rate in patients with these devices has been reduced from 25% to 2.0% to 3.5%. ICD implants also help resolve the problem of bradyarrhythmias either causing or complicating VT or VF. Timing of the implant depends on the underlying pathology and concomitant therapeutic approaches (see Chapter 12 for further details).

3. Cardiac surgery

In patients with significant CAD and surgical criteria, coronary bypass surgery decreases the incidence of sudden death when compared to patients on conventional medical treatment. The reduction is most evident in patients with 3-vessel disease, diabetes, and HF.

Surgical treatment of ventricular arrhythmias includes excision of VT foci after endocardial mapping and excision of LV aneurysms. This approach is performed with decreasing frequency because of perioperative mortality and the alternative, transvenous ICD implantation. Patients with long QT syndrome (LQTS) who do not respond to β-blockers are candidates for ICD implantation or left cardiac sympathetic denervation.

VII. DISEASES WITH INCREASED RISK OF ARRHYTHMIC SCD

A. Coronary Artery Disease

In adults, 80% of SCD episodes occur in patients with CAD. In patients with low EF, up to 50% of deaths are sudden or arrhythmic in nature. Ventricular arrhythmias leading to cardiac arrest may be due to reentry around a chronic scar or secondary to acute MI/ischemia.

No specific anatomopathologic coronary pattern indicates an increased risk for SCD: this phenomenon occurs in association with active plaque and thrombotic activity. Markers of increased risk are previous cardiac arrest, syncope, prior MI (especially within 6 months), EF < 35% (dilated cardiomyopathy [DCM] can result from prior ischemia and infarction), and history of frequent ventricular ectopy (more than 10 premature ventricular contractions (PVC)/h or NSVT).

From a therapeutic point of view, early reperfusion/revascularization is of paramount importance in reducing the risk of future cardiac events; it can be accomplished through thrombolysis, percutaneous coronary intervention, or bypass surgery. ICDs are reserved for those patients with a history of sustained ventricular arrhythmias or severe persistent LV dysfunction despite adequate medical therapy.

As LV dysfunction progresses, overall mortality increases but the percentage of sudden death decreases: this could be the result of the increasing use of β-blockers and ACEIs, drugs known to reduce the risk of SCD, achieved probably through a combination of decreased ischemia, increased VF threshold, and decreased rate of ventricular ectopy.

B. Nonischemic Dilated Cardiomyopathy

NIDCM is the presence of LV dysfunction due to nonischemic causes (viral, autoimmune, genetic, alcohol-induced, etc). It has an increasing incidence of approximately 7.5 cases per 100,000 persons per year. Among cases of SCD, 10% are estimated to be due to NIDCM. The prognosis is reserved for most of these patients, with a 5-year mortality of 20% and sudden death accounting for 8% to 51% of deaths.

The most common mechanism of cardiac arrest is ventricular tachyarrhythmia (while bradyarrhythmias and pulseless electrical activity are less frequent), likely on a substrate of ventricular fibrosis and resulting reentrant arrhythmias. Drugs used to treat HF (inotropic agents, diuretics, AADs) have proarrhythmic properties and may trigger arrhythmias in some cases.

Predictors of sudden death are previous cardiac arrest, syncope, EF < 35%, and use of inotropic medications.

Treatment should include optimal medical therapy, paired with ICD implantation in eligible patients.

C. Hypertrophic Cardiomyopathy

HCM is an autosomal-dominant, incompletely penetrant genetic disorder resulting from a mutation in one of the genes encoding proteins of the cardiac muscle sarcomere. HCM is the most common cause of SCD in people younger than 30 years, mainly athletes. SCD is often the first manifestation, typically after a vigorous exercise. Its annual mortality has been calculated between 2% and 6%.

The reason for death is usually ventricular arrhythmias, often polymorphic VT triggered by ischemia, outflow obstruction, or AF; arrhythmias may also be due to reentry, catecholamine influences, and so forth.

The electrocardiographic and hemodynamic features are not useful to predict the risk of sudden death. The electrophysiological study (EPS) also has a limited value as a risk stratifier; echocardiography, though diagnostic in many cases, sometimes does not fulfill current criteria. Genetic testing does serve to confirm the diagnosis and to characterize patients with high-risk mutations; it also aids with family screening.

The known risk factors have been categorized as "major" and "possible" in individual patients (Table 6.2).

The absence of risk factors puts the affected person in a low-risk group. The presence of multiple risk factors, especially severe septal hypertrophy, recurrent syncope, and sudden death in close relatives, warrants consideration of prophylactic ICD implantation. The presence of just one marker of high risk may justify the same

Table 6.2. Risk Factors for Sudden Death in HCM

Major	Possible
▪ Cardiac arrest (VF) ▪ Spontaneous sustained VT ▪ Family history of premature sudden death (< 40 years old) ▪ Unexplained syncope ▪ LV thickness ≥ 30 mm ▪ Abnormal exercise BP (↓>10 mm Hg or failure to ↑> 25 mm Hg) ▪ Spontaneous NSVT	▪ AF ▪ Myocardial ischemia ▪ LV outflow tract obstruction ▪ High-risk genetic mutation ▪ Intense (competitive) physical exertion

recommendation, as a significant proportion of ICD discharges have been documented in primary prevention patients with just one major risk factor.

ICDs are effective in survivors of cardiac arrest and sustained VT/VF, but in cases of primary prevention the percentage of appropriate discharges is less (5% vs. 11% per year in secondary prevention).

Selecting the appropriate type of ICD to be implanted is very important, and the decision should be taken after analyzing several conditions. A single-chamber device is probably the best option in cases of persistent AF and low probability of paroxysmal AF, and in children and adolescents; a dual-chamber ICD is a better option in the presence of LV outflow tract obstruction, bradyarrhythmias, paroxysmal AF, and neurally mediated syncope.

Patients with significant symptoms may also be treated with β-blockers or verapamil. Amiodarone is often used to control AF, which if left unmanaged often leads to significant clinical deterioration.

D. Arrhythmogenic Right Ventricular Cardiomyopathy/Dysplasia

ARVC/D is a chronic condition characterized by progressive replacement of the RV wall with fibrofatty tissue, associated with RV dysfunction and ventricular arrhythmias. Involvement of the interventricular septum and the LV is associated with poorer outcomes. In 30% to 50% of cases, it occurs as a familial disorder, with common autosomal-dominant inheritance (autosomal-recessive transmission has been reported for select mutations, such as in Naxos disease). Its prevalence is roughly estimated at 1 per 1000 individuals. Mutations in the genes encoding proteins involved in cell-to-cell adhesion (desmosomes) are found in 50% to 60% of patients with ARVC/D.

More common between 20 and 50 years of age, sudden death is a frequent first manifestation, with an annual incidence of 0.08% to 9.00%; affected patients may also complain of palpitations and syncope, often associated with exercise. Atrial arrhythmias may be present in as many as 25% of patients. Sustained monomorphic VT is common, mimicking RV outflow tract VT, but other morphologies are also possible. Often the arrhythmias are more evident than the right heart symptoms.

A typical ECG is described: precordial T-wave inversion in V_1 through V_3, QRS duration > 110 milliseconds, and presence of epsilon waves (low-voltage potentials following the QRS, characteristic but infrequent) (Figure 6.2). The signal-averaged ECG (SAECG) demonstrates presence of late potentials in 50% of patients. Modified diagnostic criteria have recently been proposed (Table 6.3), including results of molecular genetic testing.

ICDs are of utility in cases of unexplained syncope or sustained VT/VF; they should also be considered in cases of family history of sudden death and extensive disease (including LV involvement). AADs are also useful, sotalol or amiodarone,

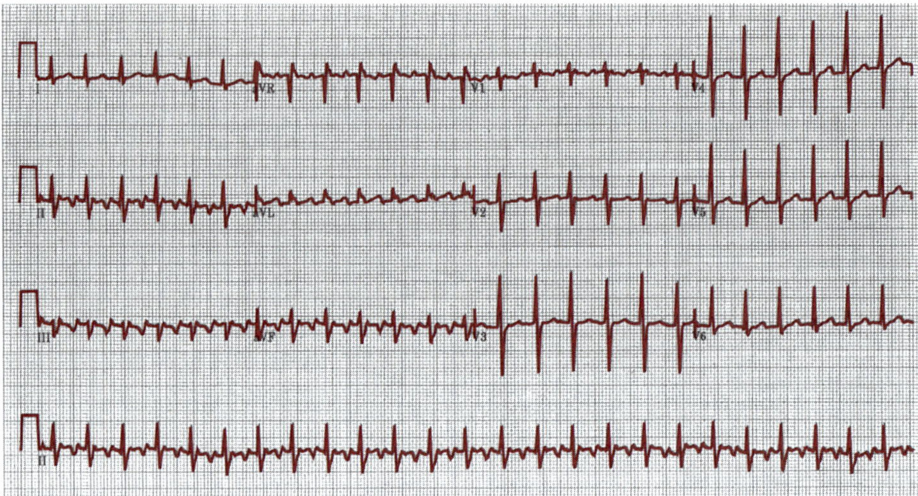

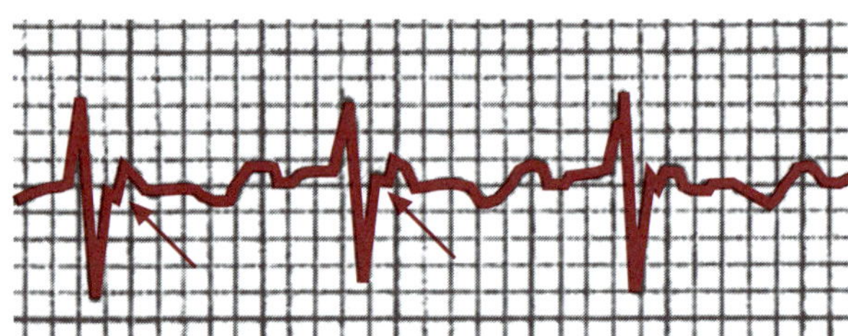

Figure 6.2. (a) ECG of a patient with ARVC/D. (b) Segment from V_1 showing epsilon waves (arrows).

alone or combined with β-blockers. Catheter ablation of symptomatic tachycardias is occasionally attempted but no clear effect in reducing sudden death has been demonstrated.

E. Long QT Syndrome

LQTS is a disorder in ventricular repolarization that leads to increased risk of life-threatening ventricular arrhythmias and SCD. Its incidence is estimated at 1 per 2000 patients. It may be acquired or congenital; both may have a genetic basis.

The common feature for all types of LQTS is the prolongation of the cardiac repolarization, predisposing to early afterdepolarizations, which in turn initiate the torsade de pointes (TdP); this trademark arrhythmia is often preceded by a pause

Table 6.3. Proposed Diagnostic Criteria for ARVC/D. Reprinted from Marcus FI, McKenna WJ, Sherrill D, et al. Diagnosis of arrhythmogenic right ventricular cardiomyopathy/dysplasia: proposed modification of the task force criteria. *Eur Heart J.* 2010;31(7):806-814, by permission of Oxford University Press.

Original task force criteria	Revised task force criteria
I. Global or regional dysfunction and structural alterations*	
Major	
Severe dilatation and reduction of RV ejection fraction with no (or only mild) LV impairment Localized RV aneurysms (akinetic or dyskinetic areas with diastolic bulging) Severe segmental dilatation of the RV	**By 2D echo:** Regional RV akinesia, dyskinesia, or aneurysm *and* 1 of the following (end diastole): — PLAX RVOT ≥32 mm (corrected for body size [PLAX/BSA] ≥19 mm/m^2) — PSAX RVOT ≥36 mm (corrected for body size [PSAX/BSA] ≥21 mm/m^2) — *or* fractional area change ≤33%
	By MRI: Regional RV akinesia or dyskinesia or dyssynchronous RV contraction *and* 1 of the following: — Ratio of RV end-diastolic volume to BSA ≥110 mL/m^2 (male) or ≥100 mL/m^2 (female) — *or* RV ejection fraction ≤40%
	By RV angiography: — Regional RV akinesia, dyskinesia, or aneurysm
Minor	
Mild global RV dilatation and/or ejection fraction reduction with normal LV Mild segmental dilatation of the RV Regional RV hypokinesia	**By 2D echo:** Regional RV akinesia or dyskinesia *and* 1 of the following (end diastole): — PLAX RVOT ≥29 to <32 mm (corrected for body size [PLAX/BSA] ≥16 to <19 mm/m^2) — PSAX RVOT ≥32 to <36 mm (corrected for body size [PSAX/BSA] ≥18 to <21 mm/m^2) — *or* fractional area change >33% to ≤40%

Table 6.3. *(Continued)*

Original task force criteria	Revised task force criteria
	By MRI: Regional RV akinesia or dyskinesia or dyssynchronous RV contraction *and* 1 of the following: — Ratio of RV end-diastolic volume to BSA ≥100 to <110 mL/m² (male) or ≥90 to <100 mL/m² (female) — *or* RV ejection fraction >40% to ≤45%
II. Tissue characterization of wall	
Major	
▪ Fibrofatty replacement of myocardium on endomyocardial biopsy	▪ Residual myocytes <60% by morphometric analysis (or <50% if estimated), with fibrous replacement of the RV free wall myocardium in ≥1 sample, with or without fatty replacement of tissue on endomyocardial biopsy
Minor	▪ Residual myocytes 60% to 75% by morphometric analysis (or 50% to 65% if estimated), with fibrous replacement of the RV free wall myocardium in ≥1 sample, with or without fatty replacement of tissue on endomyocardial biopsy
III. Repolarization abnormalities	
Major	▪ Inverted T waves in right precordial leads (V_1, V_2, and V_3) or beyond in individuals >14 years of age (in the absence of complete right bundle-branch block QRS ≥120 ms)
Minor	
▪ Inverted T waves in right precordial leads (V_2 and V_3) (people age >12 years, in absence of right bundle-branch block)	▪ Inverted T waves in leads V_1 and V_2 in individuals >14 years of age (in the absence of complete right bundle-branch block) or in V_4, V_5, or V_6 ▪ Inverted T waves in leads V_1, V_2, V_3, and V_4 in individuals >14 years of age in the presence of complete right bundle-branch block

Table 6.3. *(Continued)*

Original task force criteria	Revised task force criteria
IV. Depolarization/conduction abnormalities	
Major	
■ Epsilon waves or localized prolongation (>110 ms) of the QRS complex in right precordial leads (V_1 to V_3)	■ Epsilon wave (reproducible low-amplitude signals between end of QRS complex to onset of the T wave) in the right precordial leads (V_1 to V_3)
Minor	
■ Late potentials (SAECG)	■ Late potentials by SAECG in ≥1 of 3 parameters in the absence of a QRS duration of ≥110 ms on the standard ECG ■ Filtered QRS duration (fQRS) ≥114 ms ■ Duration of terminal QRS <40 μV (low-amplitude signal duration) ≥38 ms ■ Root-mean-square voltage of terminal 40 ms ≤20 μV ■ Terminal activation duration of QRS ≥55 ms measured from the nadir of the S wave to the end of the QRS, including R′, in V_1, V_2, or V_3, in the absence of complete right bundle-branch block
V. Arrhythmias	
Major	
	■ Nonsustained or sustained ventricular tachycardia of left bundle-branch morphology with superior axis (negative or indeterminate QRS in leads II, III, and aVF and positive in lead aVL)
Minor	
Left bundle-branch block-type ventricular tachycardia (sustained and nonsustained) (ECG, Holter, exercise)	Nonsustained or sustained ventricular tachycardia of RV outflow configuration, left bundle-branch block morphology with inferior axis (positive QRS in leads II, III, and aVF and negative in lead aVL) or of unknown axis
Frequent ventricular extrasystoles (>1000 per 24 hours) (Holter)	> 500 ventricular extrasystoles per 24 hours (Holter)

Table 6.3. *(Continued)*

Original task force criteria	Revised task force criteria
VI. Family history	
Major	
■ Familial disease confirmed at necropsy or surgery	■ ARVC/D confirmed in a first-degree relative who meets current Task Force criteria ■ ARVC/D confirmed pathologically at autopsy or surgery in a first-degree relative ■ Identification of a pathogenic mutation† categorized as associated or probably associated with ARVC/D in the patient under evaluation
Minor	
Family history of premature sudden death (<35 years of age) due to suspected ARVC/D	History of ARVC/D in a first-degree relative in whom it is not possible or practical to determine whether the family member meets current Task Force criteria
Familial history (clinical diagnosis based on present criteria)	Premature sudden death (<35 years of age) due to suspected ARVC/D in a first-degree relative
	ARVC/D confirmed pathologically or by current Task Force Criteria in second-degree relative

- PLAX indicates parasternal long-axis view; RVOT, RV outflow tract; BSA, body surface area; PSAX, parasternal short-axis view; aVF, augmented voltage unipolar left foot lead; and aVL, augmented voltage unipolar left arm lead.
- Diagnostic terminology for original criteria: This diagnosis is fulfilled by the presence of 2 major, or 1 major plus 2 minor criteria or 4 minor criteria from different groups. Diagnostic terminology for revised criteria: definite diagnosis: 2 major or 1 major and 2 minor criteria or 4 minor from different categories; borderline: 1 major and 1 minor or 3 minor criteria from different categories; possible: 1 major or 2 minor criteria from different categories.
- *Hypokinesis is not included in this or subsequent definitions of RV regional wall motion abnormalities for the proposed modified criteria.
- †A pathogenic mutation is a DNA alteration associated with ARVC/D that alters or is expected to alter the encoded protein, is unobserved or rare in a large non-ARVC/D control population, and either alters or is predicted to alter the structure or function of the protein or has demonstrated linkage to the disease phenotype in a conclusive pedigree.

Table 6.4. LQTS Diagnostic Criteria

Markers	Points
ECG	
QTc	
≥ 480 ms	3
460–470 ms	2
450 ms (in males)	1
TdP	2
TWA	1
Notched T in 3 leads	1
Low HR for age	0.5
Clinical Findings	
Syncope	
With stress	2
Without stress	1
Congenital deafness	0.5
Family History	
Family members with definite LQTS	1
Unexplained SCD in < 30 years old among immediate family members	0.5

Note: Definite LQTS, score ≥ 4 points; low probability, ≤ 1 point; intermediate, 2–3 points; high, ≥ 4 points.

("short-long-short" coupling), though it can develop without changes in HR or a preceding pause. Syncope and sudden death are typical manifestations, while palpitations are not common. Diagnostic criteria are shown in Table 6.4.

Many factors are needed to maintain normal repolarization, and there is an ongoing controversy regarding the role of simple QT prolongation as a marker of proarrhythmia susceptibility. A long QT is a common finding in the absence of drug therapy or electrolyte abnormality; when symptoms appear in this setting, it is usually in patients with genetic predisposition. On the other hand, drugs that produce the same QT prolongation do not pose an equivalent proarrhythmic risk (e.g., amiodarone vs. quinidine). A group of signs collectively known as TRIaD (Triangulation, Reverse use dependence, electrical Instability, and Dispersion) has been proposed as markers of drug-induced proarrhythmia. Amplification of transmural dispersion of repolarization has already been associated with the genesis of ventricular tachyarrhythmias in LQTS and other inherited ion channelopathies.

Table 6.5. Described Genetic Mutations in LQTS

Syndrome	Gene	Inheritance	Affected Ionic Current	Effect in Current	Symptoms	ECG
LQT1	*KCNQ1*	Dominant	I_{Ks}	↓	Exercise	Broad T
LQT2	*KCNH2*	Dominant	I_{Kr}	↓	Emotion or exercise	Bifid T
LQT3	*SCN5A*	Dominant	I_{Na}	↑	Sleep	Asymmetrical, late, peaked T
LQT4	*ANK2*	Dominant	$I_{Na, K}$ I_{NCX}	↓		Pronounced T, sinus bradycardia, PAF
LQT5	*KCNE1*	Dominant	I_{Ks}	↓		
LQT6	*KCNE2*	Dominant	I_{Kr}	↓		
LQT7	*KCNJ2*	Dominant	I_{K1}	↓		
LQT8	*CACNA1C*	De novo	$I_{Ca(L)}$	↑		
LQT9	*CAV3*		I_{Na}	↑		
LQT10	*SCN4B*		I_{Na}	↑		
JLN1	*KCNQ1*	Recessive	I_{Ks}	↓		
JLN2	*KCNE1*	Recessive	I_{Ks}	↓		

1. Acquired LQTS

Acquired LQTS usually results from electrolyte imbalance or drug therapy, often in the presence of sinus bradycardia, hypokalemia, or hypomagnesemia. Drugs are the most common cause: frequently involved are AADs (Ia and III), antipsychotics, antibiotics, diuretics, and antihistaminics. There is also an association with subarachnoid hemorrhage, nutritional deficiencies, hypothyroidism, and so forth.

2. Congenital LQTS

Congenital LQTS is a rare familial disorder; TdP is the most common arrhythmia but PVCs, monomorphic VT, bradycardia, and AV block have also been described. Mutations in several genes have already been described, leading to at least 12 different subtypes (Table 6.5). Most frequent cases are LQT1 (42%), LQT2 (45%), and LQT3 (8%), with most mutations in genes that encode for K^+ or Na^+ channels.

Phenotypically, there are 2 forms: The Jervell-Lange-Nielsen syndrome, associated with congenital deafness, has an autosomal-recessive pattern of inheritance. The Romano-Ward syndrome is not associated with deafness and has an autosomal-dominant pattern of inheritance with variable penetration (this syndrome accounts for > 90% of LQTS cases).

The clinical presentation of congenital LQTS is quite variable: many patients are asymptomatic and are referred for evaluation because of an affected family member or an incidental long QT finding. Other patients develop TdP with syncope and sudden death: exercise is a common trigger in LQT1, while events at rest or during sleep are common in LQT3.

The QT interval is best measured in leads DII and V_5 (Figure 6.3), being careful to avoid including a U wave. The upper limit of the normal for a corrected QT (QTc) is generally considered to be 440 milliseconds (450 milliseconds for adult women and 430 milliseconds for adult men). The formula most commonly used is Bazett's ($QTc = \frac{QT}{\sqrt{RR}}$), with the RR measured in seconds (the user should be aware that this formula overcorrects for HR < 60 bpm and undercorrects at high HR). A normal QTc does not exclude this diagnosis, as 10% to 20% of carriers of pathologic mutations may have a normal interval.

A paradoxical increase (≥ 30 milliseconds)in the uncorrected QT interval during infusion of low-dose epinephrine appears pathognomonic for LQT1, particularly if the ECG at rest is normal.

All symptomatic patients should be treated with β-blockers. Survivors of a cardiac arrest should receive an ICD; class 1 agents such as mexiletine may be useful in selected cases of LQT3 as adjunctive therapy.

For those patients of unknown genotype and syncope, β-blocker therapy should be instituted immediately: this therapy is highly effective in decreasing overall mortality and should be used at maximum tolerated doses. Unfortunately, β-blockers do not eliminate the risk of syncope, cardiac arrest, or SCD definitely, and they are less

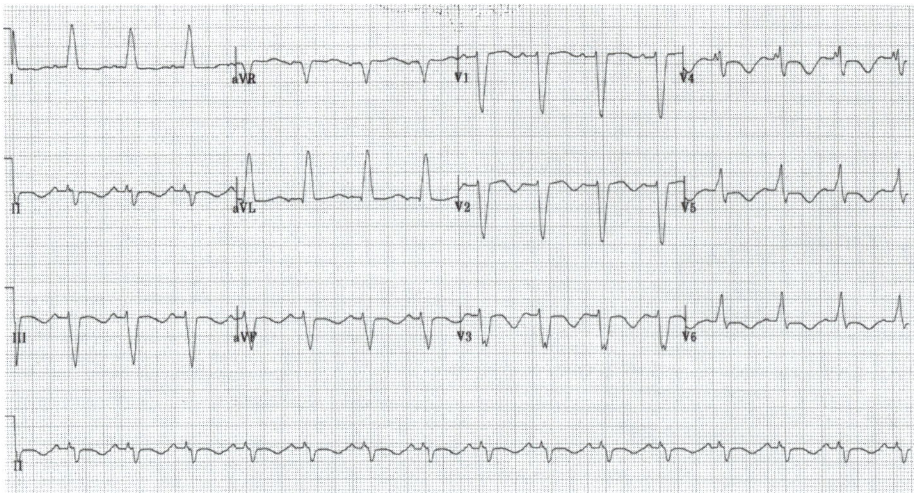

Figure 6.3. Acquired LQTS in a patient with Steinert's disease and recent use of azythromycin (QTc = 612 milliseconds).

effective in patients with mutations in Na^+ channel genes (LQT3). ICD therapy may be indicated in patients with recurrent symptoms despite treatment with β-blockers. Left sympathetic cardiac denervation has a proven antifibrillatory effect, including patients at high risk of death: it should be considered in all LQTS patients who experience syncope despite β-blockers and in those who have arrhythmia storms and shocks with an ICD. If an episode of TdP is preceded by bradycardia or pauses, permanent pacing is indicated.

Patients with known "silent" mutations have a normal QT interval due to low penetrance; their risk of cardiac events is low (< 10%) but not insignificant. These patients are mainly at risk when exposed to drugs containing an I_{Kr} blocker and should be provided with a list of potentially dangerous medications (found at www.qtdrugs.com).

When the genotype is known, gene-specific therapy is possible. LQT1 patients benefit from antiadrenergic interventions (β-blocker and sympathetic denervation), with ICDs mostly reserved for those who suffered a cardiac arrest. These patients should avoid competitive sports, especially swimming. Regarding LQT2 patients, their response to β-blockers is less effective than that of LQT1 patients and ICDs are used more liberally. Due to a proven correlation between cardiac episodes and auditory stimuli, unexpected noises should be avoided (e.g., removing alarm clocks and telephones from patients' bedrooms); K^+ levels should be always be preserved > 4 mmol/L. LQT3 patients respond less favorably to β-blockers than the other groups and management is more difficult; though not standard therapy, some groups recommend to implant an ICD even if the patient is asymptomatic (which could probably be delayed in patients < 10 years, especially girls, who are rarely symptomatic before that age). Sympathetic denervation is effective and should be offered to those unwilling to receive an ICD implant. Pharmacological therapy includes combining a β-blocker with mexiletine, but this approach requires a careful follow-up and conclusive data of its utility is lacking. LQT3 patients should also have their sleep monitored and family members trained in BLS/ACLS maneuvers.

F. Short QT Syndrome

SQTS is a newly recognized syndrome that can lead to lethal arrhythmias and SCD. Mutations in cardiac K^+ channel genes (KCNH2, SQT-1; KCNQ1, SQT-2; KCNJ2, SQT-3) lead to alterations in I_{Kr} and I_{Ks} that result in their gain of function, which causes shortening of atrial and ventricular effective refractory period (RP) and decreased QT interval.

Patients with short QT usually have a family history of sudden death. Affected individuals may complain of palpitations and/or syncope, and AF is a frequent finding. The heart is typically normal in structure. To diagnose short QT, the QTc should be ≤ 300 milliseconds, and tall and peaked T waves should be present. Extrinsic causes that may shorten the QT interval (tachycardia, hyperkalemia, hypercalcemia, digitalis intoxication, hyperthermia, acidosis) should be carefully excluded. VF is often inducible during EPS in these patients: typically the atrial and ventricular effective RPs are short, in direct relationship to the QT interval. Although some AADs (e.g., quinidine, sotalol, ibutilide, procainamide) can prolong the QT, data to support this therapy approach are still insufficient. ICD placement is the best therapeutic approach at this time, although T-wave oversensing, resulting in inappropriate ICD discharges, has been problematic.

G. Brugada Syndrome

Brugada syndrome (BrS) is an arrhythmogenic disease characterized by an ECG pattern of ST-segment elevation in right precordial leads, a structurally normal heart, and an increased risk of SCD due to VF. In 20% of affected patients a mutation in *SCN5A* on chromosome 3 is detected (the gene encoding for the α subunit of the cardiac Na^+ channel).

BrS typically manifests during adulthood, with a mean age of SCD of 41 ± 15 years. It is estimated to be responsible for at least 4% of unexpected cardiac deaths and for at least 20% of sudden deaths in patients with structurally normal hearts.

Around 20% of patients develop supraventricular tachycardias (SVTs), often AF. The most common clinical presentation is syncope. The typical ECG pattern is frequently concealed, and it is unmasked by Na^+ channel blockers, vagotonic agents, fever, α-adrenergic agonists, β-adrenergic blockers, tricyclic or tetracyclic antidepressants, hypo- and hyperkalemia, hypercalcemia, alcohol, and cocaine toxicity, and so forth.

BrS is definitely diagnosed when a type 1 ST-segment elevation (Figure 6.4) is observed in more than one right precordial lead (V_1 through V_3) in the presence or absence of a Na^+-channel-blocking agent, in conjunction with one of the following:

- Documented VF
- Polymorphic VT
- Family history of SCD at < 45 year old
- Coved-type ECGs in family members
- Inducibility of VT during EPS
- Syncope
- Nocturnal agonal respiration

The type 2 (saddleback appearance with a high takeoff ST-elevation ≥ 0.2 mV, a trough displaying ≥ 0.1 mV ST elevation, and then either positive or biphasic T waves) and type 3 (saddleback or coved appearance with ST-elevation < 0.1 mV) patterns are not diagnostic, but a positive diagnosis of BrS is also made when they turn into a type 1 ECG after the administration of a Na^+ channel blocker (e.g., procainamide) (plus one or more of the previous clinical criteria).

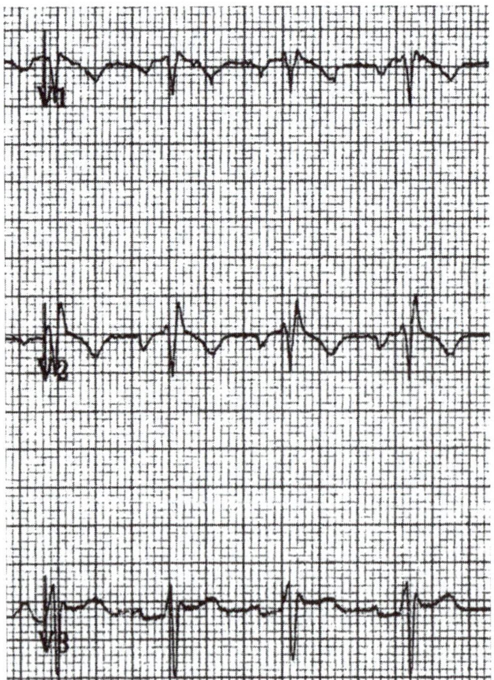

Figure 6.4. Type 1 Brugada pattern, with coving ST elevation (≥ 0.2 mV) in V_1 to V_2, downsloping ST segment and inverted T waves, pseudo RBBB pattern with no reciprocal ST changes, and normal QTc.

Table 6.6. Characteristics of ARVC/D and BrS

Characteristics	ARVC/D	BrS
Age at presentation	25–35	35–40
Natural history	SCD, HF	SCD
M/F ratio	3:1	8:1
Circumstances	Effort	Rest
Imaging	Abnormal	Usually normal
SAECG	Positive	Negative
Pathology	Abnormal	Normal
Mechanism of arrhythmia	Scar-related	Phase 2 reentry
ECG changes	Mostly fixed	Variable
AV conduction	Normal	50% abnormal PR/HV
Repolarization	Inverted T in precordial leads	High take-off ST in V_1 to V_3
Depolarization	RsR′ V_1, epsilon waves	RBBB
Atrial arrhythmias	Late (secondary)	Early (primary in 10%–25%)
Ventricular arrhythmias	Monomorphic VT/VF	Polymorphic VT/VF
AAD effect	I, II, III ↓; IV – or ↓	I, II ↓; III – or ↓; IV –
β-stimulation	↑incidence	↓incidence

A subpopulation of patients with ARVC/D has been found to have some features characteristic of BrS, but basically these 2 are distinct entities (Table 6.6).

An ICD is indicated in symptomatic patients with a spontaneous type 1 ECG. The role of EPS in asymptomatic patients is controversial. Further risk stratification has less supporting evidence, and management is usually based on expert consensus (Figures 6.5 and 6.6).

From a pharmacological point of view, quinidine and tedisamil may exert a therapeutic action because of their I_{to}-blocking properties, but appropriate clinical trials are lacking. Other possible useful agents are isoproterenol (because it boosts the L-type Ca^{2+} current) and cilostazol (a phosphodiesterase III inhibitor that augments I_{Ca} and reduces I_{to} by increasing the HR).

H. WPW Syndrome

The presence of an antegrade-conducting accessory pathway in WPW syndrome allows for AF to be conducted to the ventricles at very high rates, which can

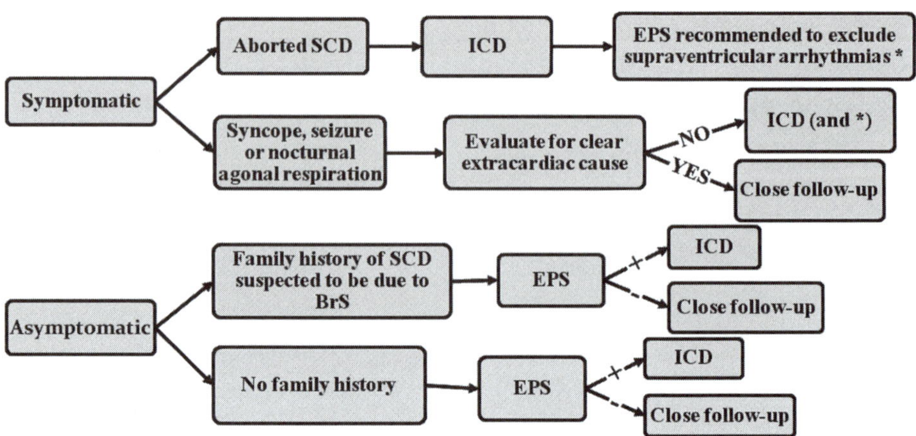

Figure 6.5. Indications for ICD implant in BrS with spontaneous type 1 ECG.

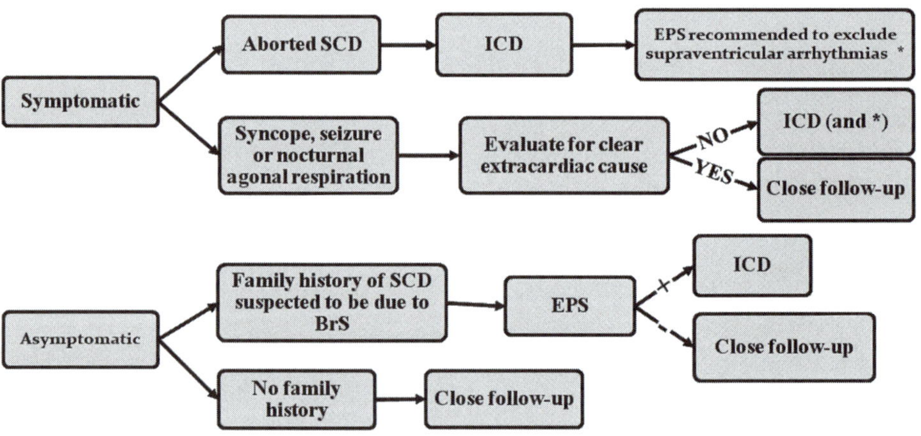

Figure 6.6. Indications for ICD implant in BrS with Na^{+-} channel-block-induced type 1 ECG.

degenerate into VF and cause sudden death. Fortunately, the overall incidence of SCD in WPW is quite low (0.15–0.39%).

Predisposing factors are a history of AF or reciprocating tachycardia; markers of higher risk are the presence of multiple accessory pathways, posteroseptal pathways, a preexcited R-R interval < 250 milliseconds during AF, and a short antegrade accessory pathway RP (< 250 milliseconds).

Electric cardioversion (ECV) should be used if hemodynamic compromise is present. Most symptomatic patients benefit from catheter ablation of the anomalous pathway or AADs (practically in disuse for this purpose nowadays). Asymptomatic patients may be assessed for markers of risk—for example, via an EST (see Chapter 9 for further details)—resulting in some patients being observed without treatment. Medications that block the AVN (digoxin, β-blockers, CCBs) are contraindicated in patients with WPW and AF because they may facilitate an accelerated conduction through the accessory pathway.

I. Catecholaminergic Polymorphic VT

Catecholaminergic polymorphic VT is a rare condition, presenting with exercise or emotional stress-induced ventricular arrhythmias, typically bidirectional VT (irregular polymorphic VT is also seen, with normal QT), often leading to sudden death. Syncope occurs in > 60% of cases during physical activity or acute emotion.

CPVT has been linked to mutations in genes for RyR2, CASQ2, and ankyrin B protein. It typically begins in childhood, and 30% of cases are familial. The resting ECG is usually normal, though sinus bradycardia, PR shortening, and prominent U waves are found in some patients. The heart is structurally normal. The medical therapy of choice is β-blockers, effective in 60% of patients: when maximal doses fail to control the arrhythmia, an ICD is indicated.

J. Primary VF

Primary VF occurs in structurally normal hearts. It is usually a diagnosis of exclusion, reached after all possible etiologies have been carefully excluded (it is likely that some earlier studies have inadvertently included patients with unrecognized BrS, LQTS, or other ion channelopathies). Most affected are typically young, with a male predominance. Even though no macroscopic cardiac abnormalities are found, there is some evidence of channelopathies and/or myocardial abnormalities, with patchy areas of activation delay during EPS. A recent study has found a positive relationship between presence of early repolarization in the ECG and occurrence of idiopathic VF.

Its only proven therapy is ICD implantation; mapping and catheter ablation of monomorphic PVCs that trigger arrhythmias is of benefit for those patients who experience frequent episodes of VF following ICD placement.

SUGGESTED READINGS

Lehnart SE, Ackerman MJ, Benson Jr DW, et al. Inherited arrhythmias: a National Heart, Lung and Blood Institute and Office of Rare Diseases workshop consensus report about the diagnosis, phenotyping, molecular mechanisms, and therapeutic approaches for primary cardiomyopathies of gene mutations affecting ion channel function. *Circulation.* 2007;116:2325-2345.

Sarkozy A, Brugada P. Sudden cardiac death and inherited arrhythmia syndromes. *J Cardiovasc Electrophysiol.* 2005;16:S8-S20.

Turakhia M, Tseng ZH. Sudden cardiac death: epidemiology, mechanisms and therapy. *Curr Prob Cardiol.* 2007;32:501-546.

Zipes DP, Camm AJ, Borggrefe M, et al. ACC/AHA/ESC 2006 guidelines for management of patients with ventricular arrhythmias and the prevention of sudden cardiac death. *J Am Coll Cardiol.* 2006;48:247-346.

CHAPTER 7

Atrial Fibrillation

Miguel A. Barrero Garcia and Peter Guerra

I. GENERAL PRINCIPLES

Atrial fibrillation (AF), defined as an supraventricular tachycardia (SVT) due to uncoordinated atrial activation with consequent deterioration in atrial mechanical function, is the most common sustained cardiac arrhythmia; it may occur in structurally normal hearts but it is most frequently seen in the presence of conditions such as hypertension, heart failure (HF), and valvular heart disease. AF can lead to impaired quality of life, thromboembolic complications, tachycardia-induced cardiomyopathy, and overall increased mortality (1.5- to 1.9-fold higher than those without it). It has an annual incidence of < 0.1% per year in patients < 40 years old and 2% in those > 80 years old. Its prevalence in the general population is 0.4% to 1.0%, increasing with age (8% in patients > 80 years).

Clinically, AF can be classified as follows:

- **First detected episode.**
- **Paroxysmal:** Self-terminating, lasting up to 7 days.
- **Persistent:** Episodes lasting more than 7 days, or that require termination, either pharmacologically or electric.
- **Permanent:** Continuous episodes, long-standing, where cardioversion has failed or was not tried.
- **Recurrent:** Two or more paroxysmal or persistent episodes.

This classification does not apply to episodes lasting < 30 seconds or due to reversible causes (cardiac surgery, pulmonary embolism, MI, alcohol ingestion, electrolyte abnormalities, myocarditis, and hyperthyroidism). When reversible causes are found, their treatment together with management of the episode of AF will often eliminate the arrhythmia and the need for long-term management.

The term *lone AF* applies to patients < 60 years old, without clinical or echocardiographic evidence of cardiopulmonary disease (including hypertension): these patients appear to have less thromboembolic risk and mortality.

II. MECHANISMS

A. Focal Origin

A focal origin of AF has been known for some time now, but recent research has provided significant clinical evidence: located mainly in the pulmonary veins (but also detected in the superior vena cava, coronary sinus, *crista terminalis*, and ligament of Marshall), the focal triggers for AF are premature potentials arising from muscular sleeves extending from the atria to the veins. The atrial tissue in these areas has distinct electrophysiological (EP) features: its refractory period (RP) is shorter than that of control patients and the atria of patients with AF; likewise, the RP is shorter in the distal pulmonary vein than at the pulmonary vein–left atrium (LA) junction.

Additionally, decremental conduction in pulmonary veins is more frequent in AF than in control patients. All these EP characteristics facilitate the rapid discharge of pulmonary vein potentials, either automatic or due to microreentry, with quick local activation of the LA, extending then less rapidly and in disorganized fashion to the right atrium (RA). Patients with focal AF usually suffer from paroxysmal AF, are typically younger, and generally have many premature atrial contractions (PACs) noted on Holter monitoring, and isolation or elimination of these foci can lead to elimination of the trigger for paroxysms of AF.

The existence of triggers for AF does not negate the importance of appropriate substrates for AF maintenance, as sustained AF in patients with focal triggers may depend on atrial fibrosis, and/or loss of atrial muscle mass with atrial dilatation, all leading to delayed interatrial conduction plus dispersion and shortening of the atrial RP.

B. Multiple-Wavelet Hypothesis

The multiple-wavelet hypothesis is an old concept, explaining AF as the result of fractionation of wavefronts propagating through the atria, resulting in self-perpetuating "daughter wavelets." This type of AF is commonly seen as a secondary arrhythmia associated with cardiac disease that affects the atria, resulting in a large atrial mass with short RP and delayed conduction. These patients tend to be older, and AF is more likely to be sustained.

AF can also be induced by other tachycardias (AVRT, AFL, AT), inflammation (either local, atrial, or systemic, typified by increased C-reactive protein), autonomic nervous system activity, atrial ischemia, atrial dilatation, anisotropic conduction, and structural changes related to aging. A genetic link has also been described, mostly in patients with lone AF.

III. CLINICAL HISTORY

Patients can be asymptomatic but many experience palpitations of variable duration, chest pain, dyspnea, fatigue, presyncope, and, less commonly, syncope. Embolic stroke or exacerbation of HF could be the initial presentation. A history of regular palpitations followed up by an irregular rhythm suggests tachycardia-induced tachycardia (e.g., AFL, AT, AVNRT, or AVRT leading to AF), but this is often difficult to diagnose based only on clinical history. The influences of the autonomic nervous system allow the characterization of a vagally mediated form of AF (episodes at night, at rest, after eating, or after drinking alcohol; history of progressive bradycardia; increased frequency with β-blockers and digitalis; good response to disopyramide and propafenone), and of an adrenergic form (provoked by exercise/emotional stress; association with polyuria; treatment of choice with β-blockers).

IV. PHYSICAL EXAMINATION

Irregularly irregular heart rate (HR) on auscultation is highly suggestive of AF, with associated signs depending on the underlying conditions and repercussions of the arrhythmia (i.e., pulmonary edema, elevated venous pressure, arterial hypotension, etc.). Signs suggestive of hyperthyroidism and valvular heart disease should be carefully sought.

V. DIAGNOSIS

A. Electrocardiogram

The ECG is characterized by irregular atrial rate > 300 bpm; no sinus P waves are recognized but irregular oscillations of the baseline (*f* waves) (Figure 7.1). Also, irregular and variable ventricular response (usually between 100 and 180 bpm is present), depending upon atrioventricular node (AVN)-conducting properties, sympathetic and parasympathetic tone, and AVN-blocking drugs used (Figure 7.2).

An AF with regular R-R intervals may be seen in patients with underlying complete AV block or ventricular-paced rhythm, or when there is interference by junctional or ventricular tachycardia (VT). A pseudo-regularization of the R-R interval may occur in patients with very short AVN RP (e.g., young patients under conditions of adrenergic stress).

Isolated wide QRS complexes during AF may be the result of aberrant conduction rather than premature ventricular contractions (PVCs), due to the so-

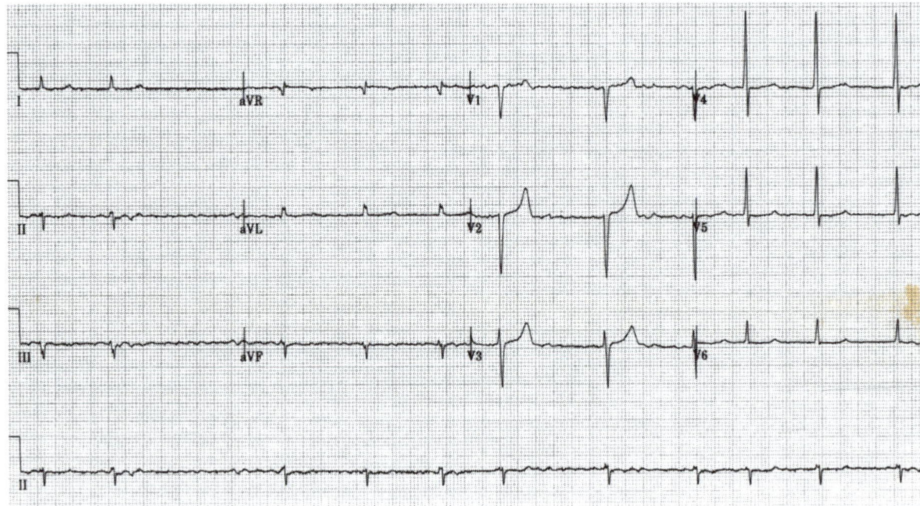

Figure 7.1. AF with a controlled ventricular response.

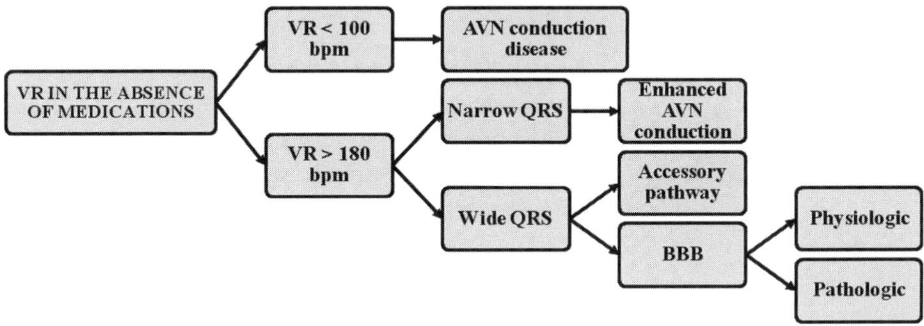

Figure 7.2. Ventricular response (VR) in AF, in the absence of AAD- or AVN-blocking medications.

called Ashman phenomenon: when a short R-R interval follows a long one, the beat of the short cycle often has right bundle branch block (RBBB) morphology. Duration of the AP (or RP) changes with the R-R interval of the preceding cycle: a longer R-R is associated with a longer duration, and vice versa; so, if a long cycle lengthens the ensuing RP, and if the following cycle is shorter, the beat ending this shorter cycle is likely to be conducted with aberrancy. When the impulse reaches the His-Purkinje system, one of its branches is still in the relative or absolute RP, resulting in conduction being slowed down or blocked in that branch (usually the right one because it has a longer RP in adults) while the beat progresses through the other one, with a BBB configuration in the ECG, all in the absence of bundle branch pathology.

In cases of rapid and wide QRS complexes a pseudo-regular AF with aberrancy or a preexcited AF are possible but a VT should always be excluded.

B. Laboratory Testing

Complete blood count, thyroid, renal, and hepatic function tests are commonly needed. Testing for digoxin and alcohol levels may be indicated; in some cases, a toxicology screen is necessary.

C. Imaging

A chest x-ray is of potential value to eliminate associated cardiopulmonary conditions. An echocardiogram is usually performed to rule out structural heart disease (SHD). A transesophageal echocardiogram is also indicated if there is need to rule out the presence of an intracardiac thrombus (e.g., AF lasting > 48 hours, or in patients with congenital heart disease [CHD]). If a pulmonary vein/LA ablation is planned, a cardiac CT scan or MRI can be of help but their use is not obligatory.

D. Electrophysiological Study

An EPS can be useful if an underlying SVT is suspected as the cause of AF. It is formally indicated if a patient presents with rapid, preexcited AF.

If pharmacologic therapy fails (either for rate or rhythm control), then an EPS and ablation (of the AVN or of the AF trigger/substrate) can be performed.

VI. THERAPY

Management of AF is addressed at controlling symptoms, either by establishing or maintaining normal sinus rhythm (NSR), or by controlling the ventricular response. Another important goal is stroke prevention.

A. Medical Management

The decision to opt for rhythm control (to restore and attempt to maintain sinus rhythm, aided if necessary by repeated cardioversion or catheter ablation) or rate control (without any specific attempt to restore or maintain sinus rhythm) should be individually tailored, as no significant difference has been found between these 2 treatment strategies, either in terms of survival rate or risk of stroke. Depending on symptoms, initial rate control could be preferred in older patients with persistent AF while younger patients, particularly if they have paroxysmal AF, should benefit from rhythm control. As a long-term goal the decision should be based not only on the clinical presentation but also on the patient's preferences, comorbid conditions, and the ongoing response to treatment. Table 7.1 lists the Canadian Cardiovascular Society recommendations for rate/rhythm control.

1. Rate control

Acute pharmacological rate control can be attained with IV β-blockers (metoprolol, propranolol, esmolol) or nondihydropyridine Ca^{2+} channel blockers (CCBs)

Table 7.1. Recommendations for Rate/Rhythm Control in AF

Favors Rate Control	Favors Rhythm Control
Persistent AF	Paroxysmal AF
Recurrent AF	First episode of AF
Less symptomatic	More symptomatic
> 65 years of age	< 65 years of age
Hypertension	No hypertension
No history of HF	History of HF
Previous AAD failure	No previous AAD failure
Patient preference	Patient preference

Table 7.2. Drugs for Acute Rate Control in AF

Drugs	Dose
Esmolol	500 μg/kg IV
Metoprolol	2.5–5.0 mg IV; repeat × 3 prn
Propranolol	0.15 mg/kg IV
Diltiazem	0.25 mg/kg IV
Verapamil	0.075–0.15 mg/kg IV
Amiodarone	150 mg IV
Digoxin	0.25 mg q2h, up to 1.5 mg

(though their use is usually associated with negative inotropy and hypotension) (Table 7.2); in patients with HF, IV digoxin or amiodarone are most useful.

When rate control is the long-term goal, a combination of digoxin with either a β-blocker or CCB is usually helpful both at rest and during exercise. Even though a HR at rest < 80 bpm (< 110 bpm during exercise) has been utilized as a marker of acceptable ventricular rate control, these HR targets should certainly be individualized as the repercussions of an inappropriate rate control are patient-specific.

2. Rhythm control

Immediate synchronized direct current cardioversion is preferred in very symptomatic patients, when the ventricular response is too rapid or in cases of hemodynamic instability. If a preexcited AF is suspected it should also be treated aggressively with immediate electric cardioversion (ECV) or IV procainamide or amiodarone (digoxin, ß-blockers, and CCBs are contraindicated in this setting).

When the duration of the AF is either > 48 hours or unknown, patients should be given IV heparin (or LMWH) and a transesophageal echocardiogram performed to rule out the presence of intracardiac thrombus, before cardioversion is attempted. If the patient is stable, rate control can be tried initially, while waiting the

Table 7.3. AAD for AF Conversion

AAD	Dose	Conversion Rate
Ibutilide	1 mg IV, repeat once prn	30%–50%
Flecainide	300–400 mg PO	50%–80%
Procainamide	15–17 mg/kg IV	30%–50%
Propafenone	600 mg PO	50%–80%
Amiodarone	150 mg IV over 10 min	30%–50%
Dofetilide	150–500 μg PO	37%–58%

Table 7.4. AAD for Maintenance of NSR in AF

AAD	Dose
Propafenone	450–900 mg PO/day
Flecainide	100–300 mg PO/day
Sotalol	80–320 mg PO/day
Amiodarone	100–400 mg PO/day
Disopyramide	400–750 mg PO/day
Dofetilide	500–1000 μg PO/day
Dronedarone	400 mg PO bid

recommended period of therapeutic anticoagulation (see following), before attempting the cardioversion. Patients who have been in AF < 48 hours should be anticoagulated depending on their thromboembolic risk.

In the case of AF scheduled for pharmacological cardioversion, flecainide, propafenone, amiodarone, dofetilide, and ibutilide are all recommended agents (Table 7.3). In cases of AF lasting > 7 days, dofetilide, ibutilide, and amiodarone are of proven benefit. It is important to note, however, that attempts at pharmacologic cardioversion do not eliminate the need to perform a transesophageal echocardiogram to rule out a LA appendage thrombus in patients at risk (see previous).

NSR can be maintained with flecainide or propafenone (in the absence of SHD) or with sotalol, amiodarone, or dronedarone (Table 7.4).

Initiation of antiarrhythmic drugs (AADs) in an outpatient setting for patients with paroxysmal AF and no associated SHD is commonly practiced; for those with significant SHD or increased risk for proarrhythmia, AADs should be initiated in a hospital setting.

Occasional breakthrough arrhythmias, without excessive symptoms, are not an indication of AAD failure and treatment should be continued; in cases of significant arrhythmia, the dose can be increased or the medication substituted.

New AADs with a promising future in controlling AF are vernakalant (atrial selective agent), azimilide (I_{Kur} blocker with use-dependent effects), and the GAP junction facilitators.

3. Stroke prevention

Patients with nonvalvular AF (absence of mitral rheumatic valve disease, mitral valve repair, or prosthetic valves) and risk factors have a 5-fold increased risk for stroke. Patients with rheumatic heart disease and AF have an even higher risk for stroke (17-fold). Anticoagulation with warfarin decreases the risk of stroke by 50% to 80%.

In the acute setting, if AF is present for < 48 hours and the patient does not have significant thromboembolic risk factors, cardioversion (either pharmacological or electric) may be accomplished safely without further need for anticoagulation (though anticoagulation therapy is recommended for all patients to minimize the risk of complications). In patients with well-tolerated AF lasting > 48 hours or of unknown duration, cardioversion is not recommended until sufficient anticoagulation is achieved. Most patients will need oral anticoagulation—for example, warfarin, with at least 4 weeks of therapeutic international normalized ratio (INR) of 2 to 3 prior to any attempt to restore sinus rhythm. Because embolic events can occur following cardioversion as atrial mechanical function returns, oral anticoagulants should be continued for an additional 4 to 6 weeks. Alternatively, a transesophageal echocardiogram may be performed initially and, if no thrombus is present, proceed to cardioversion, followed up then by therapeutic oral anticoagulation for a minimum of 4 to 6 weeks after sinus rhythm is restored.

Long-term anticoagulation is indicated in many patients. A vitamin K antagonist such as warfarin should be used in all patients at high risk of stroke (mechanical valves, prior thromboembolism, and rheumatic mitral stenosis), with target INR of 2 to 3; warfarin should also be used in the presence of more than one moderate risk factor (age ≥75, hypertension, HF, diabetes, EF < 35%). In the presence of just one risk factor, either oral anticoagulation or aspirin (81–325 mg PO od) are reasonable alternatives (though warfarin should be preferred). The same applies in the presence of one or more of the following less validated risk factors: age 64 to 74, female gender, or coronary artery disease (CAD). It is important to realize that the use of anticoagulation does not depend on the pattern of AF (paroxysmal, persistent, or permanent), as all possess similar thromboembolic risk; the exception being the patient < 60 years old, with lone AF and no risk factors.

Another method for estimating the ischemic stroke risk is the $CHADS_2$ score (Table 7.5), a more simplified version that does not take into consideration all the factors listed previously. Using this model, patients with score of 2 or greater should be anticoagulated; aspirin or warfarin (preferred) may be used when the score equals 1. Aspirin is recommended in patients with score of 0 or if warfarin is contraindicated.

Calculating the risk of bleeding (e.g., HAS-BLED score) should be part of the overall risk stratification.

Table 7.5. $CHADS_2$ Score

Clinical Condition	Points
HF	1
Hypertension	1
Age ≥ 75 years	1
Diabetes	1
Stroke or transient ischemic attack	2

Dabigatran is a direct thrombin inhibitor which may replace warfarin as the anticoagulant of choice in many cases. Several other new oral anticoagulants are currently under development.

B. Nonpharmacological Therapy

1. Complete ablation of the AVN, followed by a pacemaker implant

This procedure remains a highly effective rate-control approach in selected patients, mainly in those with tachycardia-induced cardiomyopathy, difficult control with AADs or rate-controlling agents, and no indication for other treatment strategies. It does improve symptoms and quality of life but is limited by the persistent need for anticoagulation and pacemaker dependency. During the first 1 to 3 months, the pacing rate must be programmed in the 80- to 90-beat range to prevent TdP, which is presumably due to a slow ventricular response and the occurrence of early afterdepolarizations.

2. Surgical ablation of the AF circuit

This procedure has proven to be beneficial for rhythm control but is usually limited to patients undergoing cardiac surgery for other reasons. There are 3 main types of Maze techniques: Cox-maze III, Kosakai, and Cryo-maze. Significant risk factors for a failed procedure are the presence of a large LA, small *f* waves on the ECG, and long AF duration. Minimally invasive techniques (generally with thoracoscopy) and catheter-based epicardial techniques are currently under investigation.

3. Catheter-based ablation techniques

These techniques have evolved considerably since the discovery that potentials arising in or near the ostia of pulmonary veins often trigger AF, and that their elimination abolishes AF (Figure 7.3).

In the last few years, other foci of ectopic activity triggering AF have been discovered and better mapping and ablation techniques have been put to use. Currently most operators will try elimination/isolation of all pulmonary vein potentials, or will use a nonfluoroscopic guidance system to perform an LA circumferential ablation (i.e., encircling the pulmonary vein pairs, connecting right and left pairs along the LA roof, and connection to the mitral valve annulus), with or without verification of potential elimination. There are other proposed approaches (Table 7.6).

No single ablation strategy has proved so far to be the best, and patient selection remains the key to success. In patients with paroxysmal AF and no apparent LA abnormalities, isolation of pulmonary veins is sufficient in the majority of cases. For those patients with persistent AF, with or without LA structural abnormalities, linear lesions are also required. A combined procedure is often performed, and 2 or more ablation sessions are sometimes necessary in some patients.

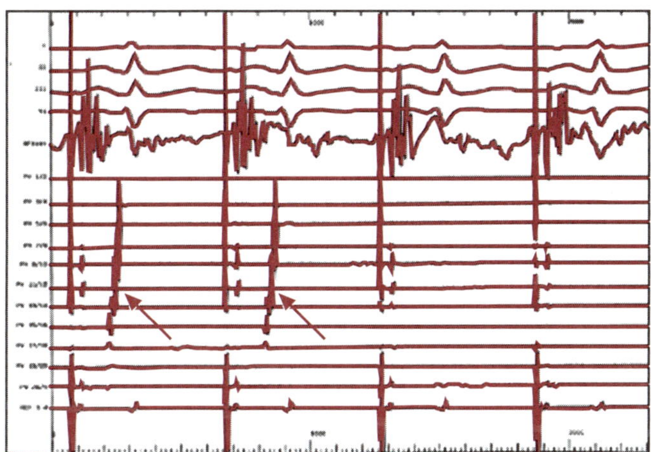

Figure 7.3. Pulmonary vein potentials within a pulmonary vein (arrows). Application of radiofrequency eliminates them.

Table 7.6. Proposed Approaches for Catheter Ablation of AF

Pulmonary vein isolation (segmental/ostial, circumferential, and circumferential/antral)
Electrogram-based ablation (CFEA ablation)
Linear ablation
Autonomic ganglionated plexi ablation
Sequential ablation strategy

The procedure is indicated mainly in cases of recurrent symptomatic AF, with failure to at least 1 to 2 different types of AAD and no significant LA abnormalities. Overall, the single-procedure success rate is about 57% for patients not on AAD and 72% for patients on AAD.

The use of chest CT scan or MRI allows the creation of a tridimensional anatomy of the LA, which aids in mapping and ablating. Intracardiac ultrasonography permits better visualization of the LA and pulmonary veins and titration of energy delivery.

Procedural complications are generally in the 5% range and include pulmonary vein stenosis, perforation, thromboembolism, cardiac tamponade, new-onset AT, and, rarely, atrioesophageal fistula. Postablation, appropriate monitoring is needed as asymptomatic episodes are often common.

The use of atrial defibrillators and atrial anti-tachycardia pacing has largely been abandoned. Surgical closing of the LA appendix (where thrombi are most commonly located) has proven efficacious in decreasing the embolic risk in selected patients; a percutaneous approach is currently under investigation.

Table 7.7. AF Outpatient Care

Regular monitoring of INR in patients on oral anticoagulation
Surveillance of drug interactions (e.g., warfarin or digoxin and amiodarone)
Monitoring of patients on AAD for signs of proarrhythmia (e.g., surveillance of QTc in patients on Ia and III drugs)
Monitoring for side effects (lung, thyroid, liver, skin) in patients on chronic amiodarone therapy
Periodic ECG and Holter monitoring to assess rate control and presence of asymptomatic episodes (with exercise testing prn)
Regular follow-up of pacemakers/ICDs

Outpatient care of AF patients is extremely important and several aspects require regular evaluation (Table 7.7). Careful monitoring for changes in HR, BP, and symptoms/signs of HF is needed. Patients on AAD should be tested regularly, looking for signs of worsening liver or renal function, which, if present, should prompt changes in the dose or in the drug.

VII. PRIMARY PREVENTION

The risk of AF may be decreased in patients with known predisposing conditions, by identifying and treating them before atrial remodeling and fibrosis take place. Well-recognized risk factors such as hypertension, diabetes, CAD, HF, and valvular heart disease should be managed comprehensively; the use of angiotensin-converting enzyme inhibitors (ACEIs) and angiotensin-receptor blockers in these conditions markedly reduces the incidence of AF, with greater benefit seen in patients with HF, LV dysfunction, and hypertension complicated by left ventricle hypertrophy (LVH). Another modifiable risk factor is excessive alcohol intake (associated with a 34% increased risk of AF). Patients with hyperthyroidism, prone to AF due to excessive sympathomimetic activity, benefit from β-blocker therapy. Targeting the renin-angiotensin-aldosterone system also appears promising, as well as managing of Ca^{2+} overload, the release of proinflammatory cytokines (typified by elevated C-reactive protein, a marker of risk even for patients without history of AF), increased oxidative stress, and the influence of protective nutritional factors such as fish consumption. Atrial or AV synchronous pacing may reduce the incidence of AF in patients with bradycardia compared to ventricular pacing. The risk of postoperative AF in patients undergoing heart surgery can also be decreased, with proven benefits obtained from perioperative use of β-blockers, sotalol, amiodarone, and atorvastatin.

VIII. CARDIOLOGY EVALUATION

Specialist advice is needed when rate/rhythm management measures fail to control the arrhythmia or when a nonpharmacological approach is indicated.

SUGGESTED READINGS

Canadian Cardiovascular Society consensus conference: 2010 atrial fibrillation guidelines update. Executive summary. www.ccs.ca. (http://www.ccsguidelineprograms.ca/index.php?option=com_content&view=article&id=112:afib-executive-summary&catid=50&Itemid=66).

Fuster V, Ryden LE, Cannom DS, et al. ACC/AHA/ESC 2006 guidelines for management of patients with atrial fibrillation—executive summary. *J Am Coll Cardiol.* 2006;48:854-906.

Natale A, Raviele A, Arentz T, et al. Venice Chart international consensus document on atrial fibrillation ablation. *J Cardiovasc Electrophysiol.* 2007;18:560-580.

Chapter 8

Atrial Flutter

Miguel A. Barrero Garcia and Laurent Macle

I. TYPICAL ATRIAL FLUTTER

A. General Principles

Atrial flutter (AFL) is a regular atrial tachycardia (AT) with a rate ≥ 240 bpm (CL ≤ 250 milliseconds, < 2% cycle-to-cycle variation) and no isoelectric baseline between the atrial deflections in the ECG. AFL has many clinical characteristics similar to atrial fibrillation (AF), but it occurs less than one-tenth as often as AF, with an overall incidence around 88 per 100,000 persons per year.

A mixture of clinical and electrophysiological (EP) characteristics are used for its classification:

1. Typical AFL

Typical AF is either counterclockwise or clockwise.

2. Atypical AFL

a. Right atrial: lower-loop reentry, upper-loop reentry, double-loop reentry, incisional macroreentry (IART), macroreentry in the free wall without history of atrial incisions, macroreentry within the superior vena cava.

b. Left atrial: LA septal circuit, mitral annulus, scar circuits.

If AFL occurs after cardiac surgery, pathogenic factors may be involved (e.g., pericarditis, atrial ischemia). AF ablation techniques, either percutaneous or surgical (e.g., MAZE-type surgery), also create a predisposing substrate that may sustain reentrant circuits (AFL, AT), mostly atypical, in a minority of patients.

The following material covers typical AFL; atypical AFL is discussed separately at the end of the chapter.

B. Mechanisms

AFL has a macroreentrant mechanism: the typical circuit is bounded anteriorly by the tricuspid annulus and posteriorly by anatomical (orifices of the superior and inferior vena cava and the Eustachian ridge) and functional (*crista terminalis*) barriers. Its inferior pivot point is the space between the inferior vena cava orifice and the tricuspid valve, the so-called cavotricuspid isthmus, a critical zone of slow conduction and the target of the ablation procedure. This circuit, seen in about 90% of clinical cases, has a counterclockwise activation sequence: when viewed in the left anterior oblique position, the wavefront proceeds in a lateral to septal direction through the cavotricuspid isthmus and then ascends along the interatrial septum, coming down along the *crista terminalis* (Figure 8.1); the opposite activation sequence is called clockwise or reversal typical flutter. Rarely, other circuits could also have cavotricuspid isthmus

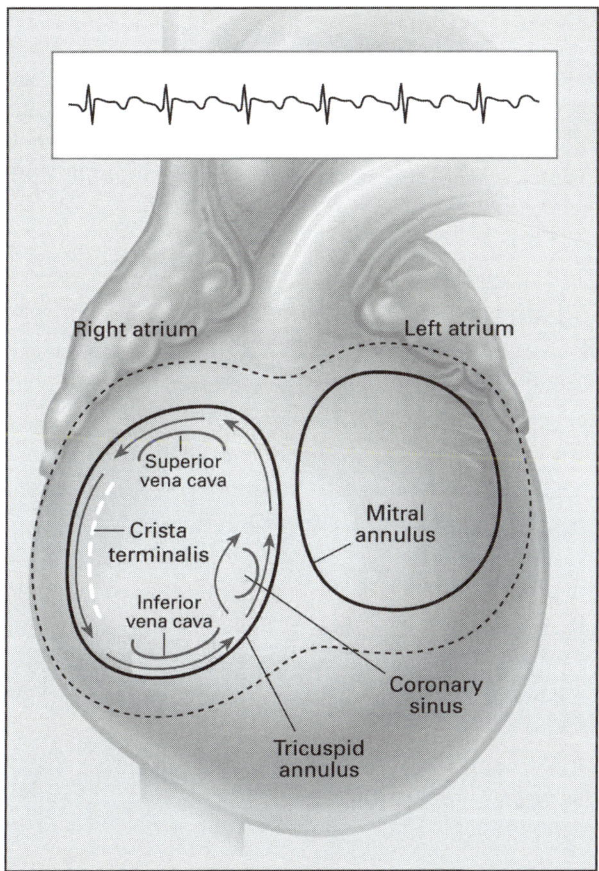

8.1. Reentry circuits in typical AFL. Reproduced with permission from Morady F. Radio-frequency ablation as treatment for cardiac arrhythmias. *N Eng J Med.* 1999;340:534-544. Copyright © 1999 Massachusetts Medical society. All rights reserved.

dependency, such as double-wave and lower-loop reentry (see section on atypical AFL for further details).

C. Clinical History

Though some are asymptomatic, most AFL patients complain of paroxysmal palpitations, with a regular rhythm, often long lasting, with spontaneous and sudden relief. Associated symptoms are chest discomfort, dyspnea, presyncope, and fatigue. Occasionally there is a history of previous heart surgery, usually for correction of congenital abnormalities, involving incision of the atria; it is noteworthy that even though this history makes an IART a strong suspect, typical circuits are still more prevalent in this group of patients (they often coexist in the same patient).

D. Physical Examination

Findings in a physical examination include the following: rapid and regular peripheral pulses; occasional cannon *a* waves, due to an atrial contraction against a closed tricuspid valve; first heart sound could be of variable intensity; hypotension, angina, or HF due to rapid ventricular response, usually in patients with LV dysfunction.

E. Diagnosis

1. Electrocardiogram

The counterclockwise circuit in typical AFL has the trademark F waves: a sawtooth pattern (downsloping segment followed by a sharper negative deflection, then a sharp positive deflection with a positive overshoot leading to the next downsloping segment), with a negative polarity in inferior leads and positive waves in V_1, transitioning to negative waves in lead V_6 (Figure 8.2). A reverse pattern is seen in cases of clockwise AFL; F waves are positive in inferior leads and negative in V_1, with transitioning to positive waves in V_6.

There is a variable ventricular response, depending on the AVN conduction properties (usually regular 4:1, 2:1). When the atrial rate is between 240 and 340 bpm it is called type I AFL; when it exceeds 340 bpm, it is known as type II AFL. Unmasking of the F waves can be obtained with vagal maneuvers (CSM, Valsalva) or rapid acting AVN-blocking agents.

Cases of 1:1 AV conduction should raise the suspicion of WPW or Lown-Ganong-Levine syndromes, phenytoine, or Ic AAD usage.

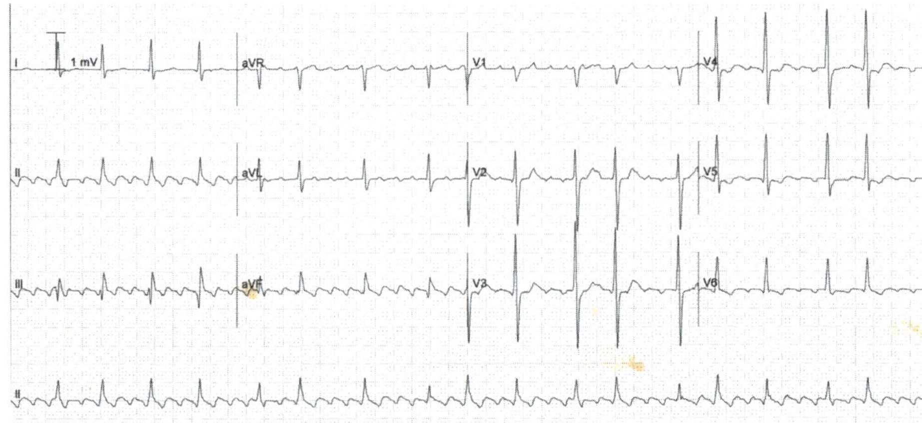

Figure 8.2. Typical counterclockwise AFL with variable AV conduction.

2. Imaging

Echocardiography is useful to evaluate cardiac function and to exclude associated structural abnormalities; a transesophageal echocardiogram helps exclude the presence of thrombus, with a similar rationale and methodology for its use as in AF.

3. EP Study

An EPS is indicated in patients who are candidates for ablation of the arrhythmia circuit. If the ECG has a persistently typical pattern (type I), ablation of the cavotricuspid isthmus can be performed without previous EP confirmation of its participation in the circuit (as this is the case in the majority of patients); clinically, cavotricuspid isthmus dependency is also suggested by persistent flutter, rather than paroxysmal, and presentation to the EP laboratory with the arrhythmia. Type II AFL usually requires EP confirmation of the cavotricuspid isthmus participation in the arrhythmia circuit, demonstrated by provoking entrainment of the arrhythmia while pacing from the isthmus (Table 8.1).

In cases of AFL with 1:1 AV conduction, the presence of an accessory pathway should always be excluded.

F. Therapy

Treatment of AFL resembles that of AF, aimed at controlling the ventricular response, rhythm management, and prevention of thromboembolic complications.

Table 8.1. EP Criteria to Establish Transient Entrainment in AFL

- Constant fusion beats in the ECG during rapid atrial pacing at a constant rate, except for the last captured beat, which is entrained but not fused (morphology of the spontaneous tachycardia)
- Progressive fusion (constant fusion beats in the ECG but with different degrees at different pacing rates)
- Demonstration that interruption of the arrhythmia is associated with localized conduction block to a site for one beat, followed by subsequent activation of that site from a different direction, manifest by a change of morphology of the EGM at the blocked site, with a shorter conduction time
- Demonstration of a change in conduction time and electrogram morphology at one recording site when pacing from another site at 2 different constant pacing rates, each of which is faster than the spontaneous rate of the tachycardia but fails to interrupt it (ECG similar to progressive fusion)

1. Rate control

Rate control in AFL is not as effective as it is in AF, and rhythm control measures should always be pursued. Rate control should be used in the following situations: while waiting for rhythm control measures to take effect, if there are no benefits in reestablishing normal sinus rhythm (NSR), if attempts to restore NSR could endanger the patient's life (significant comorbidities, or contraindications for AAD usage or ablation).

Recommended agents are verapamil, diltiazem, digoxin, and amiodarone, in doses similar to those used in AF (Table 7.2).

2. Rhythm control

a. Electric cardioversion (ECV): Very effective, and of choice if the patient is very symptomatic or with hemodynamic compromise. A synchronized 50- to 100-J biphasic shock is usually enough to stop AFL in > 90% of cases.

b. Antiarrhymic drugs (AADs): Pharmacological cardioversion can be attained with several AADs, all of which possess a variable efficacy (Table 8.2). In cases where AADs will be used on a long-term basis (less common option since the advent of catheter ablation), class Ic (of choice if no SHD), Ia or III agents are preferred. Concomitant use of an AVN-blocking agent is mandatory when AADs are used to treat AFL, in order to avoid rapid (often 1:1) AV conduction.

c. Rapid atrial pacing: Cumulative success rate is about 82%. It is a procedure of choice if the patient is in postoperatory, still with atrial epicardial wires attached; it has no place if atrial rate is > 350 bpm as the arrhythmia cannot be entrained (and therefore cannot be interrupted by pacing). The pacing should be maintained for at least 10 seconds after the P wave in lead II becomes positive. Overdrive pacing through the use of a transesophageal electrode is also effective.

Table 8.2. AAD Therapy for Acute Conversion of AFL

AAD	Conversion Rate
Flecainide: 300 mg PO single dose	13%
Propafenone: 600 mg PO single dose	40%
Ibutilide: 1 mg IV over 10 min	38%–76%
Dofetilide: 8 mcg/kg over 15 min	34%–70%
Sotalol: 1 mg/kg over 10 min	20%–40%
Amiodarone: 5 mg/kg over 10 min	22%–41%

d. Catheter ablation: Percutaneous ablation of the cavotricuspid isthmus in typical AFL (type I) is a curative and safe procedure, effective in > 90% of cases; it is the rhythm control strategy of choice and should be offered to patients in cases of significant recurrence, or even after a first significantly symptomatic episode without a trial of AADs.

e. AVN ablation, followed by pacemaker implantation: As a last resort, if catheter ablation fails to interrupt the circuit after at least a second attempt (ideally using a tridimensional mapping system), it is not possible or if the patient remains symptomatic despite AAD, he or she might be candidate for complete AVN ablation with insertion of a permanent pacemaker.

3. Stroke prevention

Thromboembolic risk in AFL is similar to that of AF, and its management does not differ.

In terms of outpatient care, use of an AAD and/or oral anticoagulation should be followed closely, as per AF. Successful ablation of type I AFL (with demonstrated bidirectional conduction block in the isthmus) results in very low recurrence rate (< 10%), with a very good prognosis. Association of AFL with AF (25%–35% of cases) is more problematic and patients will need regular assessment.

G. Cardiology Evaluation

Cardiology consult is needed when initial rate or rhythm-control measures fail, or when the patient is candidate for ablation of the arrhythmia.

II. ATYPICAL ATRIAL FLUTTER

A. Mechanisms

Atypical AFL is not dependent on conduction through the cavotricuspid isthmus and occurs in any region of the atria with conduction barriers that can support reentry, in patients with or without previous cardiac surgery or underlying SHD; double-wave and lower-loop reentry AFL are exceptions, as the isthmus is part of the circuit. In patients with CHD, recurrent atypical AFL could be an indication of deteriorating hemodynamic function.

B. Clinical History

Symptoms are similar to those of typical AFL but it tends to have more paroxysms. The hemodynamic tolerance in patients with underlying complex cardiopathies, surgically corrected or not, is usually very low (e.g., late after Senning or Fontan operations, where it becomes a marker for worse prognosis).

C. Diagnosis

1. Electrocardiogram

The atrial rate criteria apply as in typical AFL but the morphology of F waves is different, often with periodic changes (Figure 8.3). Definitive diagnosis requires intracardiac mapping, but the following features have been described in some atypical circuits:

a. RA circuit: When the circuit is located in the RA, the F waves are negative in V_1 and positive in inferior leads.

In lower-loop reentry, the ECG is suggestive of typical counterclockwise AFL due to a circuit located around the inferior vena cava, with conduction across the *crista terminalis* (the cavotricuspid isthmus is part of the circuit).

In upper-loop reentry, the circuit is confined to the upper portion of the RA, and the ECG is similar to the reverse typical AFL. A differentiation can be made observing the F waves in lead I: if they are negative, isoelectric, or flat, it is most likely upper-loop reentry; if the amplitude is > 0.07mV, then reverse typical AFL should be suspected.

b. LA circuit: LA AFL usually has flat/low-amplitude waves in inferior leads, with a positive F wave in V_1 (or positive/negative).

2. Imaging

Imaging is carried out as per typical AFL. Cardiac CT scan or MRI is often used to better define the cardiac anatomy in patients postsurgery. Angiography is commonly

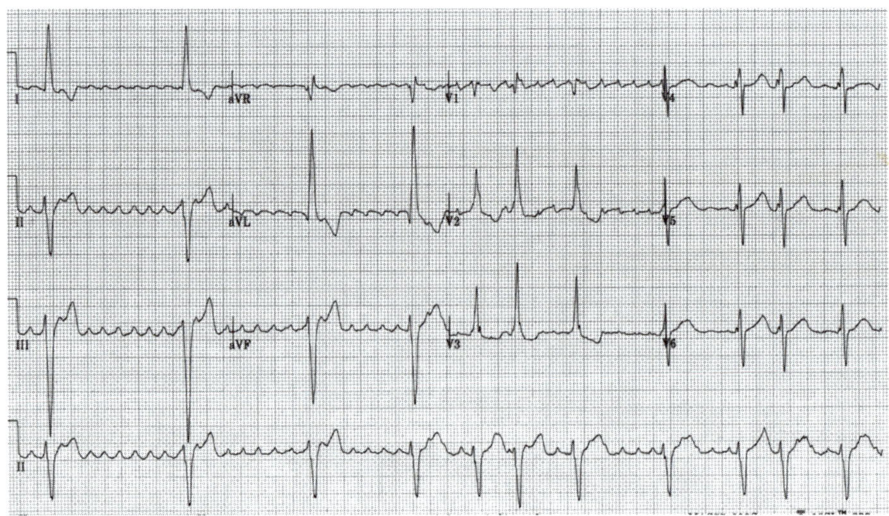

Figure 8.3. Atypical AFL with variable AV conduction (there are also a complete RBBB, a left anterior fascicular block, and LVH).

performed in the EP laboratory before the ablation takes place, helping determine the cardiac anatomy, position, and status of conduits and surgically constructed shunts, and so forth.

3. EP Study

In patients who are candidates for catheter ablation, the EPS is essential in determining the mechanism of the arrhythmia, locating the circuit, and differentiating the atypical flutter from those with isthmus dependency. A stepwise approach is required as macroreentry can occur in any region of the atria with conditions to support it (Figure 8.4); evaluation should always start in the RA, as mapping of LA circuits requires a transseptal approach.

The use of tridimensional mapping systems facilitates locating the circuit of the arrhythmia by combining pace-mapping with entrainment, substrate (voltage) mapping, and activation sequence mapping.

D. Therapy

1. Rate control

Rate control is used in similar situations as in typical AFL. Selection of drugs should be made very carefully, based on the patient's underlying condition and the drug safety profile.

2. Rhythm control

Again, ECV and AADs are used with similar rationale as in typical AFL. Rapid atrial pacing is usually ineffective as atypical flutters cannot be entrained.

Catheter ablation is the rhythm-control strategy of choice. In patients with a history of CHD, repaired or not, both typical and atypical circuits can coexist, so detailed EP mapping is needed. Success of the ablation is variable (50%–88%) and

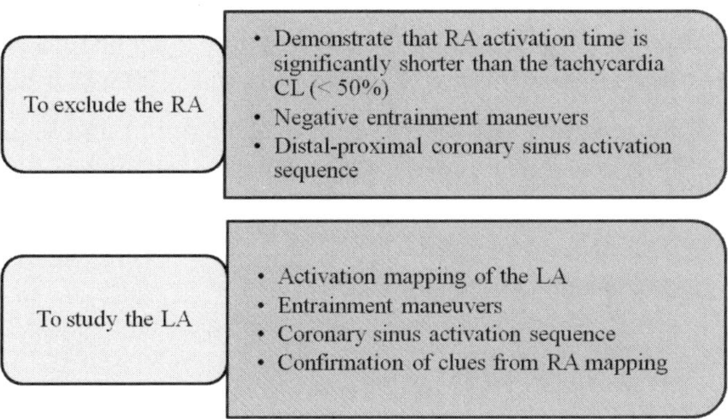

Figure 8.4. Suggested EPS approach in studying atypical AFL.

the procedure inevitably laborious; it should be performed only in centers with adequate expertise.

The treating physician should always keep in mind that an intractable tachyarrhythmia in a CHD patient could be an indication of progressive deterioration of the disease, which may warrant surgical revision of conduits and/or conversion of palliative surgical procedures to complete repairs, paired with antiarrhythmia surgery, in the hope of establishing a more physiologic circulation, stopping the arrhythmia, and preventing further cardiac deterioration.

Occasionally, complete AVN ablation followed by a permanent pacemaker implant is needed.

3. Stroke prevention

Done as per AF/typical AFL. In CHD patients an additional risk factor for thromboembolism is the implant of intracardiac leads in the presence of intracardiac septal defects.

E. Cardiology Evaluation

Atypical AFL is notorious for symptomatic recurrences and intolerance/resistance to drug therapy. Most patients benefit from specialized consult, as in many instances resolution/amelioration of symptoms (which does not necessarily imply an elimination of the arrhythmia) is achieved through nonpharmacological means.

SUGGESTED READINGS

Delacretaz E. Exploring atrial macroreentrant circuits. *J Cardiovasc Electrophysiol.* 2005;16:688-689.

Kannankeril PJ, Fish FA. Management of intra-atrial reentrant tachycardia. *Curr Opin Cardiol.* 2005;20:89-93.

Lee KW, Yang Y, Scheinman MM. Atrial flutter: a review of its history, mechanisms, clinical features, and current therapy. *Curr Prob Cardiol.* 2005;30:121-168.

Saoudi N, Cosio F, Waldo A, et al. Classification of atrial flutter and regular atrial tachycardia according to electrophysiologic mechanism and anatomic bases. *J Cardiovasc Electrophysiol.* 2001;12:852-866.

Chapter 9

Supraventricular Arrhythmias

Miguel A. Barrero Garcia and Marc Dubuc

I. GENERAL PRINCIPLES

While isolated premature contractions are usually benign, sustained supraventricular tachyarrhythmias (SVTs) are a frequent cause of morbidity in the general population. SVT is a tachycardia that emanates from or requires participation of supraventricular tissue; its incidence is about 35 cases per 100,000 patients per year, and its prevalence is about 2.25 per 1000. SVTs can either be persistent or paroxysmal and include atrial tachycardias [other than atrial fibrillation (AF) and atrial flutter (AFL)], AV nodal reentrant tachycardia (AVNRT), and AV reentrant tachycardia (AVRT).

SVTs usually have a narrow QRS complex (width < 120 milliseconds), which indicates activation via the His-Purkinje system; less commonly a wide QRS tachycardia can be of supraventricular origin in patients with abnormalities in AV conduction (either preexistent or rate-dependent) or, much less frequently, when antegrade conduction of an SVT occurs over an accessory pathway.

Reentry is the most common mechanism of narrow QRS complex tachycardia, with increased automaticity and triggered activity occurring less frequently (Figure 9.1).

Approximately 60% of cases of SVT are due to AVNRT, 30% to AVRT, and 10% to AT. When analyzing a narrow QRS tachycardia, all possible etiologies to tachycardias originating above or from within the AVN should be taken into account:

- **Sinus node-related:** sinus tachycardia, inappropriate sinus tachycardia, SNRT.
- **Atrium-related:** AT (focal or multifocal), AF, AFL.
- **AVN-related:** AVNRT, AVRT, junctional tachycardias.

Immediate treatment of a SVT depends, as is the case for ventricular arrhythmias, on the presence or not of hemodynamic compromise. Stable patients allow for thorough examination of the ECG to proper identify the rhythm disorder and select the best therapeutic approach (Figure 9.2). Continuation of the tachycardia despite some degree of AV block is highly suggestive of AT or AFL, excludes AVRT, and makes AVNRT very unlikely. When vagal maneuvers fail to stop the arrhythmia, the use of an antiarrhythmic drug (AAD) is often useful. In patients with hemodynamic instability, prompt electric cardioversion (ECV) is recommended.

This chapter focuses on the 3 most common types of SVT, with brief references to junctional tachycardia and short PR syndrome.

II. PREMATURE SUPRAVENTRICULAR CONTRACTIONS

A. General Principles

Premature contractions may arise from the atria, from a site other than the sinoatrial node (SAN), or the AVN itself (the so-called premature junctional contractions). They are common in the general population, with or without structural heart

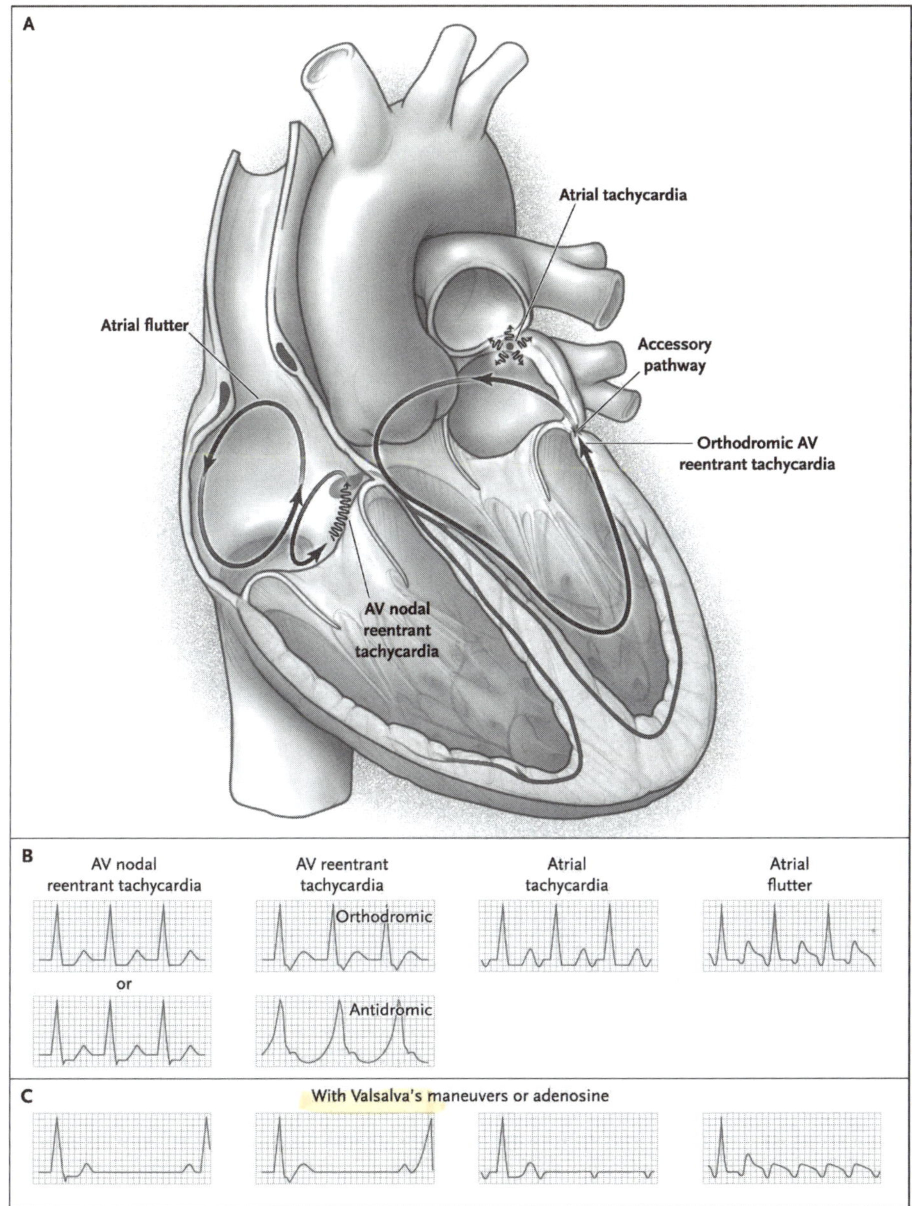

Figure 9.1. EP mechanisms in SVT. Reproduced with permission from Delacretaz E. Supraventricular tachycardia. *N Eng J Med*. 2006;354:1039-1051.

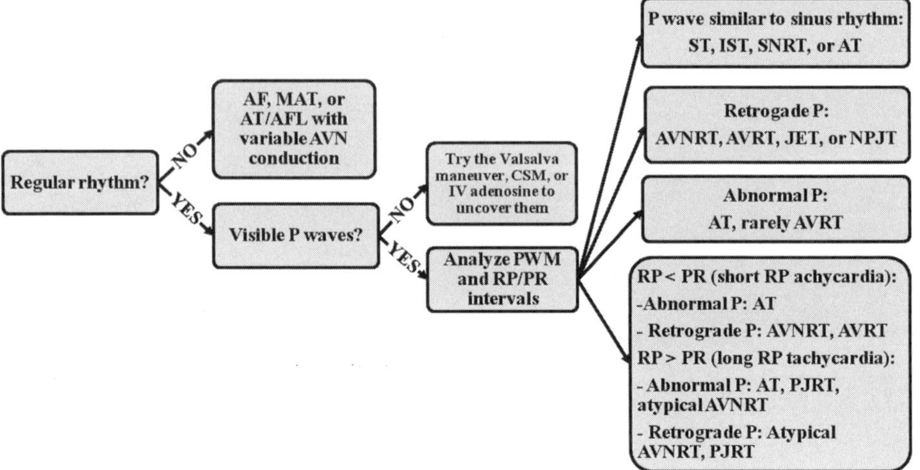

Figure 9.2. ECG analysis in narrow QRS tachycardia. JET = junctional ectopic tachycardia, abnormal P = abnormal morphology other than inverted P wave, ST = sinus tachycardia.

disease (SHD), and not necessarily an abnormal finding. They tend to be more frequent in older persons; an association between premature atrial contractions (PACs) and increase risk of stroke, hypertension, and AF has been described.

B. Mechanisms

Reentry seems to be the most common electrophysiological (EP) mechanism for PACs, but enhanced automaticity and triggered activity are also possible (see "III. Atrial Tachycardia" for further details on the EP basis of these mechanisms). Junctional premature contractions are mostly due to enhanced automaticity, with either antegrade, retrograde, or bidirectional conduction.

C. Clinical History

PACs are often associated with mitral valve prolapse and left ventricle (LV) dysfunction, as well as chronic obstructive pulmonary disease. Though some patients are asymptomatic, isolated palpitations are common; occasionally PACs serve as triggers for SVT or AF.

D. Diagnosis

1. ECG

a. PAC: Premature P waves, with different morphology from the sinus P wave; PR interval usually shorter (depending on the site of origin of the ectopic beat); variable ventricular conduction (with blocked P waves, normal QRS complexes, or aberrant AV conduction).

b. Junctional premature contractions: Premature, normal QRS complex and no visible P wave (due to masking by the QRS complex or lack of retrograde atrial activation). Occasionally a P wave preceding the QRS complex may be visible, with a PR interval too short for the atrial contraction to be conducted through the AVN (< 90 milliseconds of PR duration usually indicates a junctional rather than a conducted PAC); in cases of retrograde conduction, the P may be seen at the end of the QRS or later.

2. Holter monitoring
Indicated in symptomatic patients with no clear documentation of the arrhythmia.

3. EP Study
Not indicated.

E. Therapy

No therapy is required in asymptomatic patients. Symptomatic cases should be treated initially with reassurance and elimination of precipitating factors (e.g., caffeine, smoking, stress). If necessary, very symptomatic patients can be treated with β-blockers or CCBs; rarely AAD (mostly type III) are needed.

III. ATRIAL TACHYCARDIA

A. General Principles

AT is a usually regular cardiac arrhythmia arising from the atrium with a rate > 100 bpm (CL < 600 milliseconds), originating outside the SAN, with a focal or macroreentrant mechanism (the macroreentrant atrial tachycardias are usually considered a different entity and are discussed in Chapter 8). Occasionally an AT has more than one site of origin.

There are 3 different EP mechanisms responsible for AT:

1. Enhanced automaticity: An acceleration of the normal automatic pacemaker, mainly due to an increase in the slope of phase 4 depolarization of the AP. Shortening of the refractory period (RP) and decrease in the threshold for excitation might be associated mechanisms.

2. Triggered activity: Mediated by afterdepolarizations, low-amplitude oscillations at the end of the AP that result in repeat depolarization before normal repolarization is completed. Early afterdepolarizations occur during phase 2 or 3 of the AP while late afterdepolarizations occur during phase 4. Isolated afterdepolarizations may result in premature contractions, which, if repetitive, could lead to a sustained tachyarrhythmia.

3. Microreentry: This mechanism requires a small circuit, with conduction sufficiently slow that the tissue can recover its excitability and be reexcited by the time the wave of depolarization returns. Such slow conduction may be due to a variety of mechanisms, including a rise in the maximum diastolic potential, changes in threshold, alterations in the activation and inactivation of channels responsible for depolarization or repolarization, and poor cell-cell coupling due to a decrease in transmission through the gap junction. Microreentry is usually associated with heterogeneous electrical properties, such as those present in fibrous (scar) tissue.

B. Focal AT

1. General principles

Focal AT is characterized by a regular atrial activation from a single small atrial area with centrifugal spread, with rates usually between 100 and 250 bpm (rarely at 300 bpm). When paroxysmal and self-limited, it is referred to as repetitive focal AT (FAT); when present > 90% of the time it is called incessant AT. In symptomatic arrhythmia patients its prevalence is estimated at 0.46%. It is more frequent in patients with SHD but may also occur in the absence of heart conditions.

2. Mechanisms

FAT can result from abnormal automaticity, triggered activity, or microreentry. It originates from a well-defined small area (often the tricuspid annulus, *crista terminalis,* coronary sinus, or pulmonary veins) and then spreads centrifugally to the atria.

3. Clinical history

Symptoms can start at any age: sustained palpitations of variable duration (usually with abrupt onset and termination), chest pain, fatigue, dizziness, and syncope. When incessant, it could lead to tachycardia-induced cardiomyopathy. Nonsustained AT is fairly common in older patients.

4. Diagnostics

a. ECG: Typically, P waves are discernible at a rate of 130 to 240 bpm (could even be between 100 and 300 bpm), with a clearly defined isoelectric baseline between them. The RP interval is usually long, > 60 milliseconds (> 50% of the R-R interval), with a RP > PR interval.

P-wave morphology (PWM) is different from sinus P wave; its analysis may help determine the location of the ectopic focus. *Grosso modo,* the presence of an inferior

P-wave axis suggests an origin high in the right atrium (RA), while a negative P wave in the inferior leads suggests a low atrial focus; a negative P wave in lead I suggests an origin in the left atrium (LA).

The ventricular rate is often rapid, occasionally with 1:1 AV conduction; variability in the heart rate (HR) is nevertheless not uncommon, as maintenance of AT does not depend on AVN conduction.

QRS morphology is usually similar to that of sinus rhythm even though aberrancy might be seen at very high HR.

A differential diagnosis is to be made with other types of SVT where the RP is long (atypical AVNRT and PJRT). A variable RP relationship strongly suggests AT (RP tends to be constant in AVNRT and AVRT as both depend on AVN conduction).

The response of AT to adenosine is variable, due to overlap of the EP mechanisms. When the P waves are not quite visible, using IV adenosine (which provokes increased AV block) helps to unmask them; continuation of the arrhythmia, without modification of the atrial CL, confirms the presence of AT.

b. Imaging: Echocardiography is useful to evaluate cardiac function and to rule out structural abnormalities. Occasionally a transesophageal echocardiogram is needed to exclude the presence of thrombus, mostly in patients with known SHD and persistent arrhythmia.

c. EP Study: Indicated in patients who are candidates for catheter ablation of the arrhythmia focus/foci. Some particular EP characteristics of AT are described in Table 9.1.

5. Therapy

a. Acute episodes: Vagal maneuvers are usually unsuccessful and ECV has limited effect (automatic AT is unresponsive); the same applies to overdrive pacing. IV adenosine can be used but results are variable.

Table 9.1. EP Characteristics of FAT

- Isoproterenol is usually required to initiate it
- AT is excluded if the tachycardia terminates with AV block, or if it terminates with burst ventricular pacing that does not depolarize the atrium
- Presence of AV conduction during arrhythmia other than 1:1 (2:1, 3:1, etc.)
- With a PVC applied while the His is refractory, AT is excluded if the atrial activation is advanced (excepting the rare cases of AT with bystander accessory pathway) or if the tachycardia terminates without atrial depolarization

Rate control can be obtained with IV β-blockers (which could also terminate automatic or triggered AT) or CCBs (they frequently terminate nonautomatic AT) (Table 9.2). Class Ic AADs may be of help.

b. Long-term pharmacological control: Asymptomatic FAT, or mildly symptomatic in patients without SHD, usually does not require specific therapy.

In symptomatic patients, CCBs and β-blockers are first-choice agents, occasionally combined with digoxin (Table 9.3). Among other drugs, flecainide, sotalol, and amiodarone have shown some good results but, overall, all AADs have low efficacy.

Incessant AT is notoriously resistant to drug therapy, with some reporting success with β-blockers and flecainide; its management is primarily ablative.

A potential for thromboembolism should be carefully assessed as some patients may need long-term oral anticoagulation.

c. Nonpharmacological therapy: Catheter ablation is recommended in patients with significant symptoms, poorly responsive to medical therapy, and in cases of incessant AT (regardless of symptoms). It should be performed in experienced centers, aided if needed by noncontact mapping systems and intracardiac ultrasonography. The ablation success rate is ~86%, with recurrence rate of 8%. The incidence of significant complications is low (1%–2%) (cardiac perforation, sinus node dysfunction [SND], phrenic nerve damage, AV block); it is slightly higher for left AT due to the need of transseptal puncture.

Table 9.2. Drug Therapy for Acute Episodes of SVT

Drug	IV Dose
Adenosine	6 mg; repeat 12 mg within 1–2 min if no response
Verapamil	5 mg q3–5 min, up to 15 mg
Diltiazem	0.25 mg/kg over 2 min; if no response, additional dose of 0.35 mg/kg; maintenance infusion of 5–15 mg/h
Metoprolol	5 mg over 2 min; up to 3 doses in 15 min
Esmolol	250–500 μg/kg over 1 min; then 50–200 μg/kg/min over 4 min PRN
Propranolol	0.15 mg/kg over 2 min
Procainamide	30 mg/min to maximal dose of 17 mg/kg; maintenance 2–4 mg/min
Flecainide	2 mg/kg over 10 min
Propafenone	2 mg/kg over 10 min
Ibutilide	≥ 60 kg: 1 mg over 10 min < 60 kg: 0.01 mg/kg over 10 min Repeat once if no response within 10 min

Table 9.3. Drugs for Prophylactic Treatment of SVT

Drug	Usual Maintenance Dose
SVT without Preexcitation	
Metoprolol	50–200 mg PO daily
Bisoprolol	2.5–10.0 mg PO daily
Atenolol	50–100 mg PO daily
Diltiazem	180–360 mg PO daily
Verapamil	120–480 mg PO daily
Digoxin	0.125–0.375 mg PO daily
SVT with Preexcitation or Refractory to AVN-blocking Agents	
Flecainide	100–300 mg PO daily
Propafenone	450–900 mg PO daily
Amiodarone	200 mg PO daily
Sotalol	160–320 mg PO daily

6. Cardiology evaluation

Suggested for symptomatic patients, intolerant/resistant to medical therapy. Incessant AT should also be referred for specialized care.

C. Multifocal AT

1. General principles

Multifocal AT (MAT) is an irregular tachycardia characterized by 3 or more different PWMs at different rates, most commonly associated with underlying pulmonary (~60% of cases) and/or cardiac diseases. Patients affected are usually old (~70 years) and seriously ill. MAT is considered a marker of poor prognosis in patients with acute exacerbations of chronic obstructive pulmonary disease.

2. Mechanisms

MAT has multiple foci located in different atrial locations, likely due to enhanced automaticity from a hyperadrenergic state or triggered activity. A single focus with different exit pathways or abnormalities in intraatrial conduction could also produce a similar picture.

3. Clinical history

Apart from symptoms and signs due to the underlying disease, the rapid rate usually leads to further aggravation of cardiopulmonary failure and hemodynamic deterioration.

4. Diagnostics

a. ECG: Atrial rate > 100 bpm; P waves with at least 3 different morphologies (best seen in leads II, III, and V1), separated by isoelectric intervals; variable PP intervals, PR duration, and R-R intervals.

If there are changing PWMs but the rate is 60 to 100 bpm the arrhythmia is then called wandering atrial pacemaker; the term *multifocal atrial bradycardia* is used when the rate is < 60 bpm.

MAT is the only tachycardia, other than AF, that could present with an irregularly irregular rhythm.

5. Therapy

Management of MAT is addressed at reversing the precipitating cause and treatment of the underlying condition. Occasionally medications like theophylline could induce or exacerbate MAT.

Supplementation with Mg^+ and K^+ may control MAT. If it persists despite it, careful use of verapamil (as first choice), esmolol, or metoprolol may be of help. Standard AADs are usually inefficacious.

Complete ablation of the AVN with a pacemaker implant is an option in symptomatic patients with very rapid ventricular response and resistance to pharmacological therapy.

6. Cardiology evaluation

Advisable for symptomatic patients intolerant/resistant to medical therapy, or when a nonpharmacological approach is planned.

IV. JUNCTIONAL TACHYCARDIA

Junctional tachycardia originates from the AV junction at a rate > 100 bpm and may be dissociated from the atrium. A slower variant is the accelerated junctional rhythm, with a HR of 60 to 100 bpm (certainly not a true tachycardia due to its HR < 100 bpm). Other recognized subsets are:

- **Nonparoxysmal junctional tachycardia:** A usually benign tachycardia with rates 60 to 120 bpm and a typical "warm-up" and "cool-down" pattern. In children it is most commonly seen after repair of congenital heart defects, whereas in adults it relates to conditions such as digitalis toxicity, acute MI, and recent valve surgery. Treatment is aimed at correcting the underlying abnormality.
- **Congenital junctional ectopic tachycardia:** Observed exclusively in the pediatric population. The HR tends to be very fast and may lead to tachycardia-induced cardiomyopathy if not recognized early and treated adequately. AADs are not very efficacious (propafenone has been used suc-

cessfully) and catheter ablation could be necessary. A transient junctional ectopic tachycardia is often seen after cardiac surgery in children.

- **Focal junctional tachycardia:** A rare, highly symptomatic tachycardia found in young adults, often exercise-related, with rates of 110 to 250 bpm. It may cause tachycardia-related cardiomyopathy. Patients tend to respond well to β-blockade but catheter ablation may be needed, though the risk of AV block is high (5%–10%).

V. ATRIOVENTRICULAR NODAL REENTRANT TACHYCARDIA

A. General Principles

AVNRT is a regular SVT that results from reentry within the AVN and/or perinodal atrial tissue. It is the most common form of paroxysmal SVT (two-thirds of cases), and the majority of affected patients have structurally normal hearts.

B. Mechanisms

AVNRT has a reentrant mechanism, based on the concept of dual AVN physiology (Figure 9.3): the fast pathway (rapid conduction, longer RP) is based in the compact node, while the posterior nodal extension seems to host the slow pathway (slower conduction and shorter RP), with both pathways joining in the lower common pathway.

In the typical form of AVNRT (80%–90% of cases), or "slow-fast" (Figure 9.4), a PAC arrives at the AVN when the fast pathway is still in its RP; therefore, the impulse may travel down the slow pathway, through the lower common pathway, to the bundle of His. If the fast pathway has recovered its excitability by the time the impulse reaches the final portion of the lower common pathway, it can then be activated

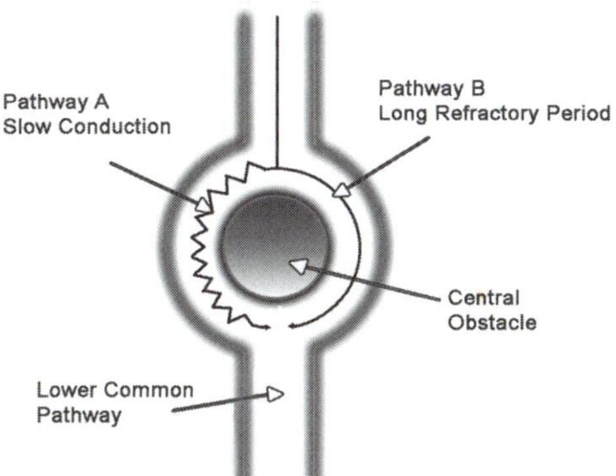

Figure 9.3. Functional anatomy of the AVN.

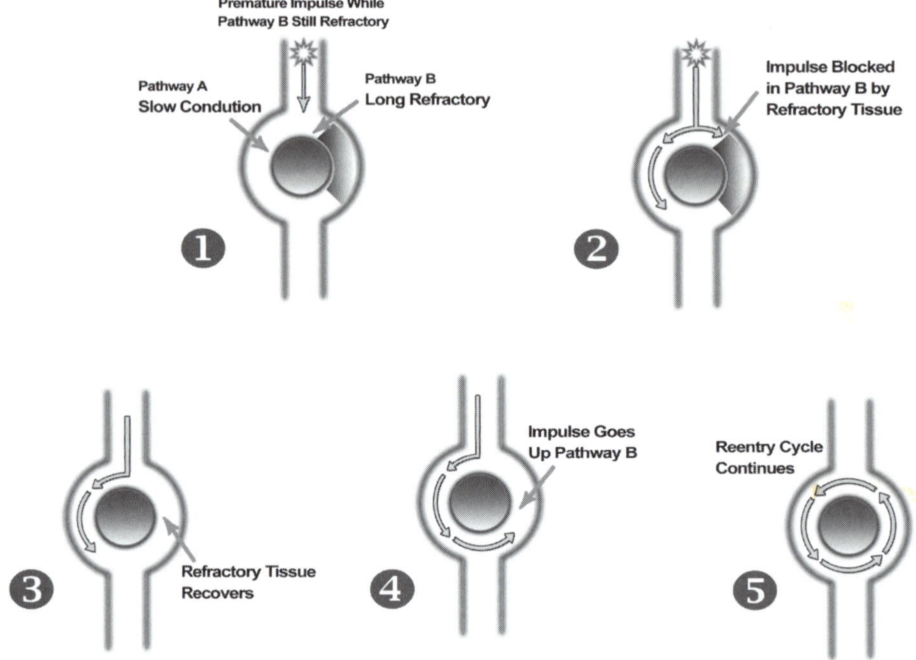

Figure 9.4. Mechanism of AVNRT. All anatomic structures are represented schematically.

retrogradely, establishing a circuit with antegrade conduction down the slow pathway and retrograde activation up the fast pathway, which results in sustained tachycardia.

Atypical forms of AVNRT can be diagnosed in up to 20% of affected patients: in the "fast-slow" variant, antegrade conduction goes down the fast pathway while retrograde conduction goes up the slow pathway. Rarely, a "slow-fast" AVNRT may be located on the left side of the heart; some patients may also have a "slow-slow" arrhythmia (antegrade and retrograde conductions both through slow nodal pathways).

C. Clinical History

Most patients are female young adults, usually with no known precipitating factors (though stress, alcohol, or exercise may trigger the arrhythmia). Regular palpitations are the most common complaint, with a sudden onset and termination. Possible associated symptoms are dizziness, dyspnea, and chest pain but they mostly depend on the HR and hemodynamic tolerance. Polyuria is commonly reported.

D. Diagnosis

1. ECG (Figure 9.5)

ECG diagnosis includes 1:1 AV conduction, with HR usually 140 to 250 bpm as well as P waves fused with the QRS; when seen at the end of the QRS (always RP < PR interval) they form a pseudo-r` in V_1 or a pseudo-S in inferior leads.

In atypical AVNRT the P waves are visible at the end of or just after the T waves (long RP tachycardia) and have a negative polarity in inferior leads. In these cases of long RP tachycardia, differential diagnosis should be made with an AT arising in the posteroseptal region and permanent junctional reciprocating tachycardia (PJRT).

2. EP Study

In patients who are candidates for catheter ablation, the EPS helps to establish the mechanism of the arrhythmia and identify the site for ablation. EP features of typical AVNRT are described in Table 9.4.

In patients with atypical AVNRT, extreme care should be exercised to avoid confusing the arrhythmia with a posteroseptal accessory pathway (Table 9.5).

Para-hisian pacing is needed sometimes, helping identify retrograde conduction over septal and right free-wall accessory pathways (but AVN conduction may mask accessory pathways located far from the pacing site or with a long retrograde conduction time).

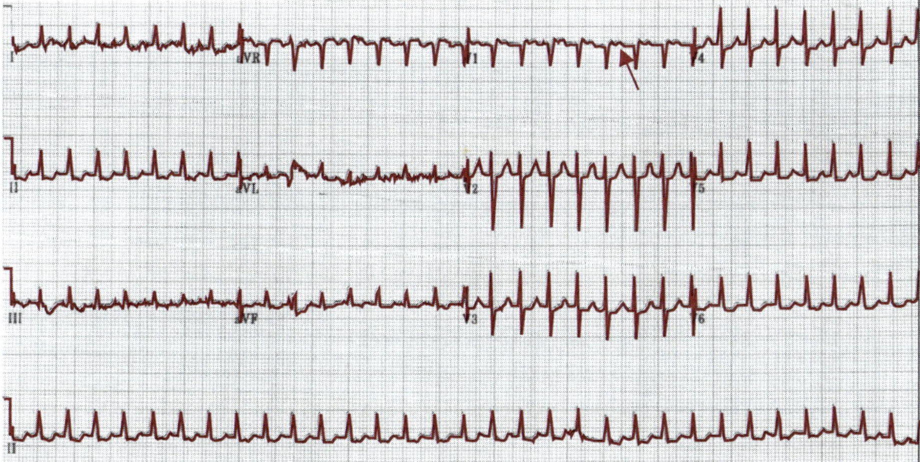

Figure 9.5. AVNRT. P waves (arrow) are seen at the end of the QRS complex.

Table 9.4. EP Characteristics of Typical AVNRT

- Progressive AH prolongation with shortening coupling interval of atrial extrastimuli (A_1A_2) until reaching a critical point where abrupt prolongation > 50 ms ("jump") in A_2H_2 occurs, followed by reentrant atrial echo/echoes or tachycardia
- Concentric retrograde atrial activation sequence (earliest at His catheter)
- VA interval during tachycardia < 60 ms at the His recording site
- PVC during tachycardia, delivered when His is refractory, should not advance the atrial activation; if earlier PVC advances His, the atrial activation should be advanced with the same activation sequence

Table 9.5. EP Characteristics of Atypical AVNRT

- Eccentric retrograde atrial activation sequence (earliest at electrodes 9–10 of coronary sinus)
- Long VA interval during tachycardia > 60 ms at the proximal His recording site
- Retrograde atrial activation sequence on the coronary sinus catheter usually is not constant during tachycardia or ventricular pacing but changes occasionally
- Activation at high RA earlier than expected if it were an accessory pathway (VA interval probably < 90 ms)
- PVC during tachycardia, delivered when His is refractory, should not advance the atrial activation; if earlier PVC advances the His, the atrial activation should be advanced with the same activation sequence

E. Therapy

1. Medical management

a. Acute episodes: Vagal maneuvers (Valsalva, CSM) may terminate the tachycardia. Due to its rapid effects, good tolerability, and high conversion rate, IV adenosine is frequently used to stop acute episodes (it also has a very good diagnostic value). Control in highly symptomatic but hemodynamically stable patients can be attained with digoxin, CCBs, β-blockers, or Ic AADs (see Table 9.2).

Patients with hemodynamic compromise can be treated successfully with ECV.

b. Long-term control: Mildly symptomatic patients with short-lasting episodes or with episodes that respond well to vagal maneuvers usually do not require specific therapy.

CCBs, β-blockers, and Ic AADs are useful to control more symptomatic patients (see Table 9.3) but long-term pharmacological therapy is in disuse since the advent of catheter ablation.

2. Nonpharmacological therapy

Percutaneous catheter ablation is recommended as first-line therapy in patients with significant symptoms, based on a very high success rate (close to 100%) and a low risk of complications.

Once the EP mechanism has been established, localization of slow pathway potentials within the triangle of Koch (which is demarcated by the tendon of Todaro, the septal leaflet of the tricuspid valve, and the orifice of the coronary sinus) usually indicates the site for effective ablation of the slow pathway, with subsequent interruption and noninducibility of the arrhythmia. Slow-fast AVNRT has a left variant, very rare, with successful ablation performed within the coronary sinus or at the posteroseptal area of the mitral annulus. Ablating the slow pathway in atypical AVNRT also terminates the arrhythmia.

The risk of inadvertent complete AV block when radiofrequency energy is used is ~1%, but with cryoablation it is almost nonexistent. The tachycardia may recur in 3% to 7% of patients, and a second procedure may be needed.

F. Cardiology Evaluation

Recommended in patients with significant symptoms, as a nonpharmacological approach is available with a good risk-benefit ratio.

VI. ATRIOVENTRICULAR REENTRANT TACHYCARDIA

A. General Principles

AVRT is a reentrant arrhythmia whose circuit involves the atrium, the AVN, the ventricles, and one or more accessory AV connections (located anywhere along the left or right AV rings). It is not common in the general population (a delta wave is detectable in 0.15%–0.25%) but constitutes a large proportion of patients suffering from SVTs and referred for EP assessment.

Accessory pathways may conduct impulses in either the antegrade (atrioventricular [AV]) or retrograde (ventriculoatrial [VA]) direction or both; only those with antegrade conduction will show a delta wave on the ECG, which is the surface indication of ventricular preexcitation. The association of preexcitation with SVT is termed WPW syndrome, but accessory pathways capable of only retrograde conduction also sustain AVRT, though no delta waves are ever seen (concealed accessory pathway). Occasionally an accessory pathway may act just as a "bystander" for an SVT, which means it is not necessary for the initiation and maintenance of the arrhythmia.

B. Mechanisms

The AP in accessory pathways depends mainly on rapid inward Na^+ currents, which allow for their typical features: very fast conduction, usually nondecremental (there is no delay in accessory pathway conduction in response to increased pacing rates), and rapid recovery of excitability (due to a short RP).

Based on EP characteristics, several types of accessory pathways and resulting arrhythmias can be identified.

1. Orthodromic AVRT

During tachycardia the reentrant impulse conducts through the accessory pathway from the ventricle to the atrium (retrograde direction) and over the AVN and His-Purkinje system from the atrium to the ventricle (antegrade conduction). During sinus rhythm the accessory pathways may be manifest (showing preexcitation on the ECG) or concealed. The tachycardia always has a narrow QRS complex, unless aberrant conduction is present.

2. Antidromic AVRT

Antegrade conduction during tachycardia occurs through the accessory pathway, from the atrium to the ventricle, while retrograde conduction goes over the AVN. Only 5% to 10% of patients with WPW syndrome present this type of AVRT and the resulting QRS is wide, making it difficult to distinguish from other types of preexcited tachycardias (such as those due to bystander accessory pathways or secondary to 2 separate conducting accessory pathways in the antegrade and retrograde portions of the circuit) and VT.

3. Bystander accessory pathway

These accessory pathways are not a critical part of an arrhythmia circuit but serve only as a bystander to the arrhythmia—for example, AT, AFL, AVNRT, or AF with preexcitation. As a consequence of antegrade conduction over the accessory pathway, the resulting QRS complexes are wide, as in antidromic AVRT.

4. Mahaim fibers

These accessory pathways, most commonly atriofascicular, have tissue properties similar to normal AVN tissue. They exhibit decremental antegrade conduction but do not conduct retrogradely; therefore, the tachycardia usually has an antidromic mechanism (preexcited tachycardia). Because the Mahaim fibers terminate in the ventricles into or near the conducting system, no delta waves are usually seen during sinus rhythm.

5. PJRT

In this type of AVRT the accessory pathway has a slow concealed conduction, only in the retrograde direction; it is located commonly in the posteroseptal region (at the ostium of the coronary sinus in > 80% of cases). This decrementally conducting accessory connection is usually sensitive to adenosine.

C. Clinical History

Palpitations usually start during adolescence. HR tends to be faster than in AVNRT, so more patients are symptomatic, with dizziness, dyspnea, chest pain, and presyncope/syncope.

PJRT is most common in children and tends to be incessant (but alternating with sinus rhythm for short periods of time). Tachycardia-induced cardiomyopathy is not uncommon.

D. Diagnosis

1. ECG

In sinus rhythm the delta wave is the hallmark of ventricular preexcitation, associated with a short PR interval and wide QRS complex (Figure 9.6a). The documented tachycardia usually has a narrow QRS, which implies an orthodromic mechanism or (more rarely) wide complexes (due to antidromic mechanism, preexcited tachycardia, or aberration), with a ventricular rate in the 150- to 250-bpm range (Figure 9.6b).

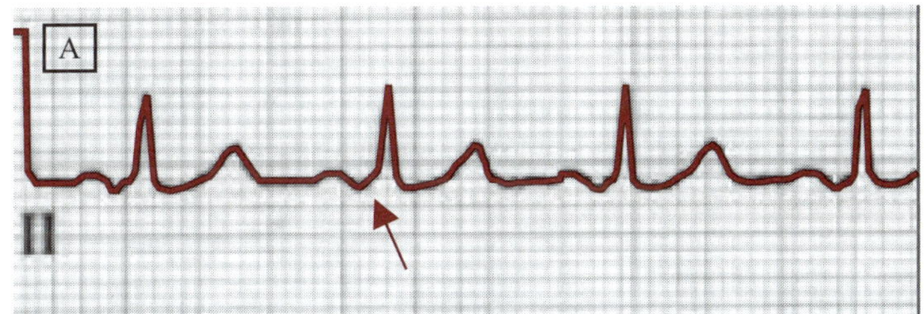

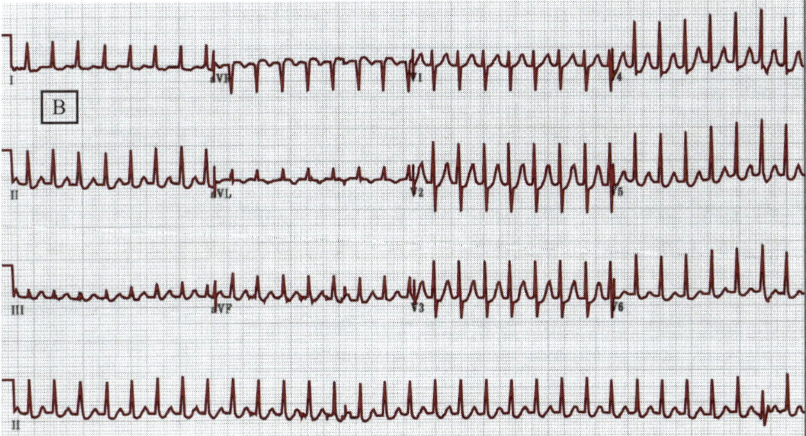

Figure 9.6. (a) Ventricular preexcitation (the arrow indicates the delta wave). (b) Orthodromic AVRT in the same patient.

QRS alternans and ST segment depression are fairly common findings (both due to the fast HR).

During orthodromic AVRT, the ECG will show P waves inscribed within the ST to T-wave segment with a short RP (usually less than one-half the tachycardia R-R interval); this RP remains constant, regardless of the tachycardia CL.

Depending on the pattern, it is possible to determine the general location of the accessory pathway (Figure 9.7).

Conduction over atriofascicular pathways (Mahaim) during sinus rhythm has variable features, often intermittent: QRS axis between 0° and –75°; QRS width ≤ 0.15 seconds; R wave in lead I; rS pattern in lead III; rS in V_1, transition from a predominantly positive QRS complex > V_4.

During tachycardia, Mahaim patients typically show a LBBB-like QRS complex with left axis deviation.

PJRT is characterized by a long RP interval with inverted P waves in II, III, and aVF.

2. EP Study

The most common EP features of accessory pathways are exposed in Table 9.6.

Right anteroseptal pathways have earliest retrograde activation at the His recording site and para-hisian pacing is needed to establish the diagnosis.

Mahaim pathways have particular EP characteristics (Table 9.7).

Because PJRT tends to be sensitive to adenosine, using this drug is not helpful in distinguishing the arrhythmia from atypical AVNRT.

If a preexcited tachycardia is suspected, the most useful maneuver is to introduce an atrial extrastimulus late enough not to affect the normal His activation: if preexcitation of ventricular activation occurs with the same sequence then there is antegrade conduction down an accessory pathway.

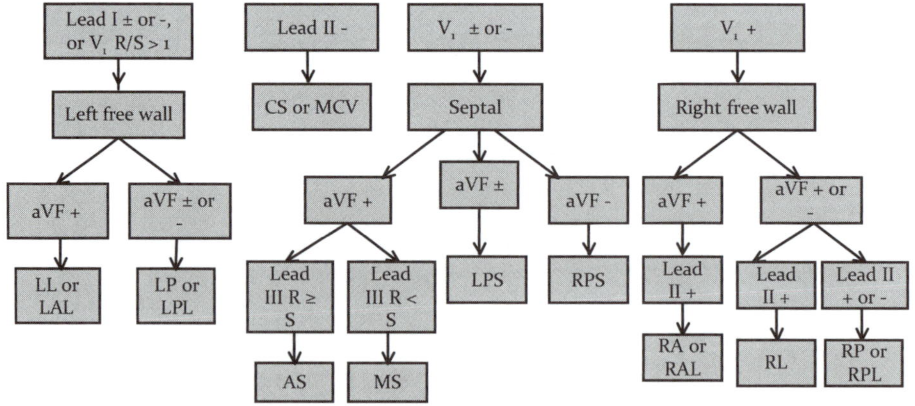

Figure 9.7. ECG algorithm for identifying accessory pathway location. LL = left lateral, LAL = left anterolateral, CS = coronary sinus, MCV = middle cardiac vein, AS = anteroseptal, MS = midseptal, LPS = left posteroseptal, RPS = right posteroseptal, RA = right anterior, RL = right lateral, RP = right posterior, RPL = right posterolateral.

Table 9.6. EP Characteristics of AVRT

- Eccentric retrograde atrial activation sequence
- Nondecremental retrograde conduction over the accessory pathway (excepting Mahaim and PJRT)
- VA conduction time shortens as the pacing site moves from the apex toward the base of the RV (highly suggestive of posteroseptal pathway)
- Earlier retrograde atrial activation along the coronary sinus catheter in left-sided accessory pathways
- Increase in tachycardia CL due to an increase in VA conduction time with functional BBB indicates an ipsilateral accessory pathway (excepting midseptal pathways)
- PVC during tachycardia, delivered when His is refractory, advances the atrial activation with the same sequence and resets the tachycardia (a changed atrial sequence indicates a bystander pathway)
- After successful ablation, incremental pacing gets blocked in the AVN or IV adenosine provokes complete AVB (most accessory pathways are insensitive to adenosine)

Table 9.7. EP Characteristics of Atriofascicular Pathways

- Prolongation of AV interval with LBBB pattern during tachycardia and atrial pacing, with superior axis
- The ventricular EGM during tachycardia is not preceded by a His EGM
- More preexcitation during RA than LA pacing
- Decremental anterograde conduction
- Absent retrograde conduction over accessory pathway
- Earliest ventricular activation at RV apex and earliest retrograde atrial activation in the His region
- Late atrial extrastimuli delivered at the tricuspid annulus will preexcite the ventricle with the same ventricular activation sequence
- Late ventricular extrastimuli that do not affect His activation will not advance the atrial activation or reset the tachycardia (earlier stimuli will advance the atrial activation)
- Frequent association with other accessory pathways and dual AVN pathways

E. Therapy

1. Medical management

a. Asymptomatic patients: Regular monitoring seems appropriate for asymptomatic patients with ventricular preexcitation, as the risk of sudden death is very low (0.15%–0.39% per patient-year). However, noninvasive testing may be carried out to better risk-stratify the patient: intermittent preexcitation on serial ECGs and sudden disappearance of the delta wave during exercise stress testing (EST) have been found reliable markers of low risk. Markers of higher risk are the presence of multiple accessory pathways, preexcited R-R interval < 250 milliseconds during AF and a short (< 270 milliseconds) antegrade accessory pathway RP: any of these may warrant prophylactic catheter ablation, indicated mostly in fully informed patients willing to undergo the procedure as well as in those patients with high-risk occupations (airline pilots, firefighters, competitive athletes).

In patients with suspected right-sided pathways an echocardiogram may be indicated to exclude Ebstein's anomaly.

b. Symptomatic patients: Acute episodes are treated as per AVNRT.
For long-term control: Mildly symptomatic patients with short-lasting episodes or with episodes that respond well to vagal maneuvers usually do not require specific therapy. Sporadic episodes lasting longer (> 1 hour) may benefit from the "pill in the pocket" approach (single oral dose of verapamil, flecainide, or propafenone). More symptomatic patients without preexcitation can be treated with CCBs, β-blockers, or digoxin, alone or in combination (see Table 9.3). It is important to remember that verapamil and digoxin are contraindicated in WPW syndrome due to the risk of rapid ventricular response, causing ventricular fibrillation (VF) in patients with AF. When these medications fail to control the arrhythmia and the patient does not accept catheter ablation, Ic and III AADs may be tried.

2. Nonpharmacological therapy

Percutaneous catheter ablation is recommended in patients with recurrent significant symptoms, lesser symptoms but preferring to avoid drugs, intolerance to drug therapy, syncope, WPW syndrome, and asymptomatic preexcitation in high-risk occupations.

The targets for the ablation procedure are the site of earliest ventricular activation (ventricular insertion site of antegradely conducting accessory pathways) or the site of earliest atrial activation (atrial insertion site of retrogradely conducting accessory pathways, either during tachycardia or during ventricular stimulation); these sites correspond with the shortest AV and VA intervals, respectively, and ideally with a QS morphology of the local unipolar potential. Targeting of accessory pathway potentials is also used but is time-consuming and not necessary for successful ablation. Mapping and ablation of Mahaim fibers is more difficult; the recording of Mahaim potentials at the tricuspid annulus is a good predictor of a successful ablation site.

In earlier studies (done with no catheter temperature control) the reported initial rate of success is > 90% for most pathway locations, but recurrence of conduction happened in as many as 15% of cases with anteroseptal and posteroseptal pathways and 21% for right free-wall locations. The use of temperature-controlled techniques, specially shaped sheaths, and alternative energy sources (e.g., cryothermy) has decreased the recurrence rate to < 5%. Procedure-related complications may occur in ≤ 2% to 3% of patients: arterial damage, bleeding, arteriovenous fistula, venous thrombosis, pulmonary embolism, myocardial perforation, valvular damage, and so forth. The risk of inadvertent complete AV block during ablation of septal pathways is ~ 1% (less when using cryothermy). Cardiac tamponade occurs in 0.13% to 1.1% and death in 0.0% to 0.2%.

F. Cardiology Evaluation

Recommended in the following situations: tachycardia with a wide QRS complex, ventricular preexcitation (with or without SVT), and SVT in a patient with syncope or severe symptoms, drug resistance, or intolerance, or if the patient prefers not to use drug therapy.

VII. SHORT PR SYNDROME

A. General Principles

Occasionally an ECG shows a short PR interval (< 120 milliseconds) in the absence of evident ventricular preexcitation (no delta wave, normal width of the QRS complex). This finding is termed *short PR syndrome* and it is thought the result of enhanced conduction through the AVN or due to intranodal or paranodal accessory pathways that bypass a portion or all of the AVN.

Absence of tachycardia episodes in the presence of a short PR interval usually indicates an enhanced AVN conduction, and it may be considered a variant of the normal; when palpitations are present, the syndrome is then termed *Lown-Ganong-Levine* and is usually classified as part of the preexcitation syndromes (together with WPW and Mahaim-type preexcitation). Incidence of Lown-Ganong-Levine syndrome is about 1 per 50,000 persons, and no study has associated it with increased incidence of sudden death or decreased survival. Its diagnosis does not imply the presence of accessory pathways, as enhanced AVN conduction may be present as well.

B. Mechanisms

Lown-Ganong-Levine syndrome has been attributed to conduction through accessory pathways such as James fibers (which connect the upper portion of the AVN to its lower portion, or to the bundle of His) or Brechenmacher fibers (an atrium-His [AH] bundle), or even to the presence of a hypoplastic or small AVN. Anatomic documentation of these structural abnormalities has been achieved, but there remain doubts about their functionality and conductive properties. Enhanced AVN conduction does not seem to exist as a phenomenon separate from normal AVN physiology (though genetic mutations in genes such as *PRKAG2* may explain it). Because modern EP techniques have usually found suspected cases of Lown-Ganong-Levine as having a different arrhythmic mechanism (AVNRT, concealed accessory pathway) it seems that this syndrome is just an epiphenomenon, with a short PR interval due to accelerated (but normal) AVN conduction and palpitations originating on a different arrhythmogenic mechanism.

C. Diagnostics

1. ECG

During sinus rhythm, the ECG shows a short PR interval and narrow QRS, with no delta wave; during tachycardia there are usually narrow QRS complexes (suggesting an antidromic mechanism).

Another significant difference from WPW is the rare occurrence of AF/AFL.

2. Labs

Electrolyte and TSH levels are recommended.

Occasionally Holter monitoring or external or implantable loop recorders are necessary to document the rhythm disorder.

3. EP Study

EPS is not necessary in most cases. Only the presence of recurrent symptomatic palpitations in a patient with a short PR interval may warrant its indication, to exclude an accessory pathway.

Enhanced AVN conduction is characterized by a short AH interval, 1:1 AVN conduction at rates at high as 200 bpm, and an abnormally small increase in AH interval as atrial pacing rate is increased.

D. Therapy

Patients with a short RP interval and no history of palpitations do not need further workup/management. In cases of palpitations/documented tachycardia, management should be pursued as in other SVTs.

SUGGESTED READINGS

Blomstrom-Lundqvist C, Scheinman MM, Aliot EM, et al. ACC/AHA/ESC guidelines for the management of patients with supraventricular arrhythmias—executive summary. *J Am Coll Cardiol.* 2003;42:1493-1531.

Calkins H, Yong P, Miller JM, et al. Catheter ablation of accessory pathways, atrioventricular nodal reentrant tachycardia, and the atrioventricular junction: final results of a prospective, multicenter clinical trial. *Circulation.* 1999;99:262-270.

Delacretaz E. Supraventricular tachycardia. *N Eng J Med.* 2006;354:1039-1051.

CHAPTER 10

Ventricular Arrhythmias

Miguel A. Barrero Garcia and Léna Rivard

I. GENERAL PRINCIPLES

Rhythm disorders arising from the ventricles affect a large portion of the population; the classic dictum of correlating severity of a ventricular arrhythmia with the presence of structural heart disease (SHD) can no longer be held, as potentially deadly arrhythmias are occasionally diagnosed in patients with structurally normal hearts.

A. Classification of Ventricular Arrhythmias by ECG

Ventricular arrhythmias are classified by ECG as follows: premature ventricular contraction, accelerated idioventricular rhythm, ventricular tachycardia (NSVT [monomorphic or polymorphic], sustained VT [monomorphic or polymorphic], TdP, bidirectional VT, BBR), ventricular flutter, and ventricular fibrillation.

B. Mechanisms

A ventricular arrhythmia may be due to:

1. Reentry

Provoked by cyclic activation around a conduction block, which can be determined anatomically (a scar, a physiologic structure, or between fascicles) or functionally (e.g., PVC, idiopathic LV VT, BBR, and scar-related VT, ischemic or nonischemic).

2. Abnormal automaticity

Due to abnormal acceleration of phase 4 of the action potential (AP) (e.g., PVC, acute ischemia VT, VT/VF associated with electrolyte and acid-base disorders, myocarditis, hypoxemia, or increased sympathetic tone such as with cocaine intoxication).

3. Triggered activity

Due to impulse generation caused by a second depolarization (the afterdepolarization).

a. Favored by slow rate: Early afterdepolarizations occurring during normal repolarization (affecting phase 3 of AP): for example, TdP; polymorphic VT.

b. Catecholaminergic-dependent: Late afterdepolarizations occurring after normal repolarization (phase 4 of the AP)—for example, right ventricular outflow tract (RVOT) tachycardia; idiopathic right VT.

Certain ECG clues may help determine the specific mechanism involved. The ECG in sinus rhythm of patients with automatic or triggered tachycardias is usually normal, while an infarct pattern indicates reentry. During tachycardia, outflow tract morphology suggests triggered activity; a regular rate indicates reentry (2 different VT morphologies are not uncommon) while automaticity usually leads to irregularity. VT due to digoxin toxicity or that terminates with adenosine is typical of triggered activity.

II. PREMATURE VENTRICULAR CONTRACTION

A. General Principles

A premature ventricular contraction is a depolarization of ventricular tissue that occurs with a coupling interval shorter than that resulting from the intrinsic normal sinus rhythm (NSR). Its significance depends on the underlying heart condition; when they lead to VT/VF, they are common mechanisms for sudden cardiac death (SCD). In asymptomatic patients, 2 or more consecutive premature ventricular contractions (PVCs), or > 10% of all ventricular depolarizations during exercise stress testing (EST), increase by 2.5 times the risk of cardiovascular death.

PVCs can be classified in many different ways:

1. According to clinical significance

Benign; malignant.

2. Associated heart disease

None (idiopathic); SHD is present.

3. According to frequency

Occasional: < 10/h or < 6/min; frequent: ≥ 10/h (by Holter) or > 6/min.

4. According to their relationship to normal beats

Bigeminy: a PVC alternating with a normal beat. Trigeminy: a PVC occurring every third beat. Quadrigeminy: a PVC occurring every fourth beat. Couplet: 2 consecutive PVCs.

5. According to origin

a. Foci/morphology: Unifocal/monomorphic: The PVC originates from one focus, usually with a fixed coupling interval, suggesting reentry or triggered activity. If the morphology is similar but the coupling interval changes, the PVCs are probably not unifocal and parasystole should be excluded. *Multifocal/polymorphic:* PVCs have more than one morphology, and they may originate from more than one site; the coupling interval is variable.

b. Site of origin: LV: PVC with RBBB morphology (in case of previous MI, a RBBB morphology indicates a septal origin). RV: PVC with LBBB morphology.

The Lown's classification was widely used in the past but has been almost abandoned in clinical practice today, as no direct measure of its clinical significance, or of the AAD effect on the arrhythmia, has been found.

B. Clinical History

Patients may complain of palpitations (with sensation of a pause rather than extra beat), dizziness, chest pain, or fatigue; presyncope/syncope are rare symptoms.

C. Physical Examination

Irregular heart beat; varying intensity of heart sounds; cannon *a* waves.

D. Diagnosis

1. ECG (Figure 10.1)

Diagnosis by ECG includes wide and bizarre QRS complex; no preceding P waves; abnormal repolarization, with T wave opposite to the QRS complex direction; fully compensatory pause—the R-R interval containing the PVC equals 2 times the basic R-R (because the PVC does not alter the sinus rhythm).

Electrocardiographically, PVCs should be differentiated from:

- **Aberrant PACs:** An ectopic P wave leads to a narrower QRS complex; the compensatory pause is incomplete.
- **Fusion beats:** Present when a PVC occurs simultaneously with a conducted sinus beat, the resulting QRS complex having an intermediate morphology. The R-R interval does not change but repolarization is completely abnormal (in pseudofusion the repolarization is not significantly affected, with the T wave in same direction as the R wave).
- **Premature junctional contractions:** There are no preceding P waves; the QRS is narrower, frequently with postcomplex negative P waves in inferior leads due to retrograde atrial activation.

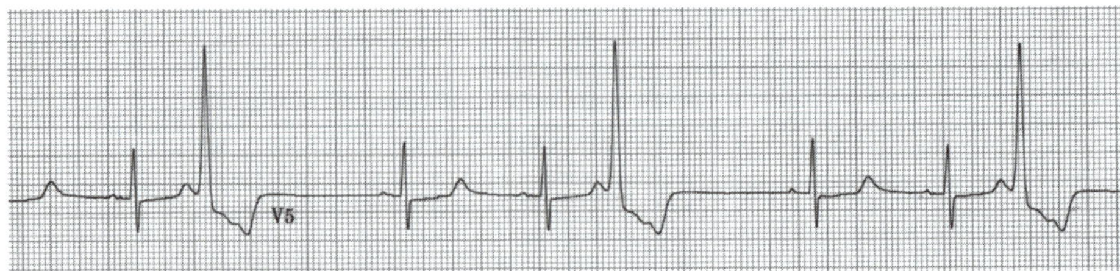

Figure 10.1. Ventricular trigeminy.

- **Idioventricular escape rhythm:** A succession of ventricular discharges at < 45 bpm, present when primary and secondary pacemaker cells fail to function.
- **Parasystole:** Occurs when a protected ectopic focus discharges independently of the dominant pacemaker. The parasystoles have wide QRS complexes: they are regular among them but with a variable coupling interval with regards to the sinus beat. Fusion beats are also found.

Holter monitoring helps characterize the PVCs and guide therapy.

2. Laboratory testing

Check for possible causes: electrolyte imbalance, medications, infections, and ischemia.

3. Imaging

Echocardiography aids in excluding SHD.

4. Exercise stress testing

EST helps to exclude exercise-induced arrhythmias or ischemia.

E. Therapy

Not all patients presenting with PVCs need specific therapy; treatment depends on symptoms, frequency, underlying condition, and risk of SCD. Antiarrhythmic agents do not improve survival.

1. Absence of SHD

Asymptomatic patients do not need specific therapy. Symptomatic patients need education and reassurance, avoidance of aggravating factors (e.g., stress, caffeine), pharmacological therapy if needed (anxiolytics, β-blockers, or CCBs).

Amiodarone has been used when other medications have failed to control very symptomatic PVCs.

2. Presence of SHD

A comprehensive evaluation is in order, including appropriate risk stratification by noninvasive or invasive means (echocardiography, nuclear imaging, Holter monitoring, SAECG, TWA, coronariography, etc). The suppression of PVCs in patients postinfarction is no longer pursued; current evidence supports the use of β-blockers but amiodarone is also safe (but it does not reduce total mortality). Hypertensive patients with left ventricle hypertrophy (LVH) should have their blood pressure (BP) appropriately controlled. Patients with low ejection fraction (EF) or at high risk for SCD benefit from an implantable cardioverter defibrillator (ICD). AADs (amiodarone, sotalol, dofetilide) should be used with caution.

Catheter ablation may be tried for PVCs that are frequent, symptomatic, monomorphic, and refractory to medical therapy, or if the patient refuses long-term medical therapy. Other possible indications for ablation are incessant VT that is consistently provoked by monomorphic PVCs and multiple ICD shocks.

F. Cardiology Evaluation

Cardiac evaluation should be considered in cases of SHD and frequent PVCs, for the purpose of risk stratification for SCD, and in patients with structurally normal hearts, when AADs or ablation is indicated—for example, RVOT PVC/VT.

III. ACCELERATED IDIOVENTRICULAR RHYTHM

A. General Principles

Accelerated idioventricular rhythm is an ectopic ventricular rhythm with 3 or more consecutive PVCs occurring at a rate < 100 bpm (CL > 600 milliseconds) but faster than the normal ventricular intrinsic escape rate of 30 to 40 bpm (CL 2000–1500 milliseconds). It is most commonly diagnosed in acute care during cardiac monitoring and its most common etiologies are myocardial ischemia, especially inferior wall ischemia/infarction; reperfusion of an occluded coronary artery; digoxin toxicity; electrolyte imbalance (e.g., hypokalemia); hypoxemia; and spinal anesthesia.

Accelerated idioventricular rhythm is nonparoxysmal and occasionally alternates control of the rhythm with NSR. In acute MI it has traditionally been considered a specific marker of successful reperfusion, but in patients treated with percutaneous coronary intervention, it has not been proven to differentiate between complete and incomplete reperfusion.

B. Mechanisms

The most likely EP mechanism is enhanced automaticity, usually associated with sinus bradycardia, which accentuates phase 4 depolarization in subordinate pacemaker cells.

C. Clinical History

The clinical history is that of the underlying disorder. Digitalis therapy should always be excluded.

D. Physical Examination

Physical examination shows occasional cannon *a* waves and varying intensity of heart sounds due to AV dissociation.

E. Diagnosis

1. ECG

Ventricular rate 60 to 100 bpm (occasionally up to 120 bpm), usually within 10 to 15 beats of the sinus rate, of gradual onset and usually well tolerated; isorhythmic dissociation with long runs of fusion beats (though 1:1 retrograde atrial capture may occur); occasional alternance with sinus rate.

There is a pathognomonic ECG pattern: a shortening of the PR interval (that occurs as the PP interval prolongs) leads to emergence of the QRS complexes; after a brief period of arrhythmia the PP intervals shorten, P waves reappear, and the SAN takes command of the rhythm.

2. Laboratory testing

Check for possible causes: ischemia, electrolyte imbalance, hyperdigoxinemia.

F. Therapy

Specific therapy is rarely required; if necessary, it can be treated with IV atropine, isoproterenol, or temporary atrial or ventricular pacing. Treatment may be considered in the presence of significant symptoms of AV dissociation , if the arrhythmia is too rapid or symptomatic; or if it is initiated by PVC with short coupling interval (R-on-T phenomenon), due to risk of inducing VF.

G. Cardiology Evaluation

Most patients with accelerated idioventricular rhythm would have already been under specialist care; if not, evaluation may be necessary when specific therapy is considered.

IV. VENTRICULAR TACHYCARDIA

A. General Principles

Ventricular tachycardia is a cardiac arrhythmia originating from the ventricles, with duration of 3 or more consecutive complexes at a rate > 100 bpm (CL < 600 milliseconds). Apart from its rate and CL, VT should always be characterized in terms of morphology, axis, hemodynamic effect, and absence or presence of SHD.

A monomorphic VT has a uniform and stable QRS configuration from beat to beat (though some variability in QRS morphology on initiation is not uncommon), while polymorphic VT has a continuously changing QRS morphology from beat to beat, with variable R-R intervals (indicating a changing ventricular activation sequence).

Occasionally, a wide QRS tachycardia (QRS ≥ 120 milliseconds) is documented but the ventricular origin is uncertain; a differential diagnosis should then

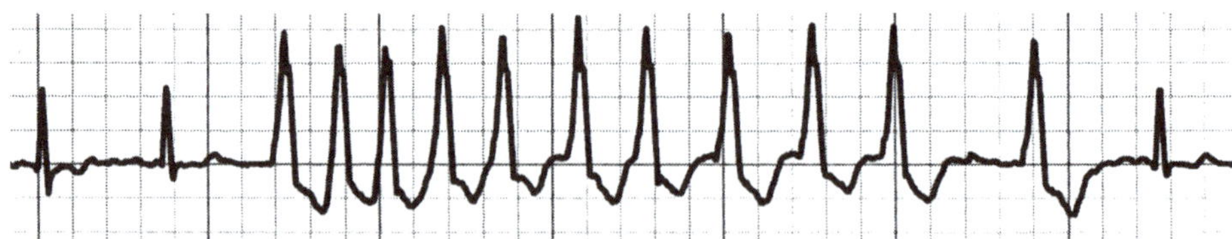

Figure 10.2. AF with aberrant AV conduction.

be established with SVT with aberrant conduction, rate-dependent (physiologic) or pathologic (preexistent BBB); ventricular preexcitation; and aberrance due to AAD use (e.g., class I agents).

When the tachycardia is irregular, the arrhythmias to be excluded are AF and AFL with variable AV conduction, in the presence of ventricular preexcitation or aberrant conduction (Figure 10.2).

The differential diagnosis of a regular wide QRS tachycardia is more difficult: the classic criterion of hemodynamic tolerance is not helpful as some patients tolerate VT fairly well. It does help to know the history of the patient (previous infarction or SHD makes VT more likely), checking previous ECGs (to exclude preexistent conduction abnormalities) and current medications. Vagal maneuvers could help unmask a supraventricular origin.

Several ECG features have been found indicative of VT: AV dissociation (identifiable in around 25% of cases); capture or fusion beats; QRS > 140 milliseconds in the absence of AAD usage; LBBB morphology with right axis deviation; concordance in the precordial leads, positive or negative (negative concordance is pathognomonic but positive concordance can also be seen with left accessory pathway); and Q peak to S interval > 100 milliseconds.

Regarding regularity, variability of up to 60 to 80 milliseconds can be seen, greatest at the onset of VT with slow rate. The Brugada diagnostic algorithm is presented in Figure 10.3.

It is noteworthy that these criteria are not helpful in cases of supraventricular arrhythmias in the presence of Ia or Ic AADs; the patient's history, vagal maneuvers, and administration of IV adenosine should help make the appropriate diagnosis. When wide QRS tachycardia is the documented rhythm, it should be treated as VT until proven otherwise.

In VT leading to out-of-hospital cardiac arrest, the use of bystander resuscitation maneuvers, automated external defibrillators, and adequate early protocols of life support help increase the survival probabilities.

B. Nonsustained VT

NSVT is defined as 3 or more consecutive PVCs at a rate > 100 bpm with a duration < 30 seconds (Figure 10.4).

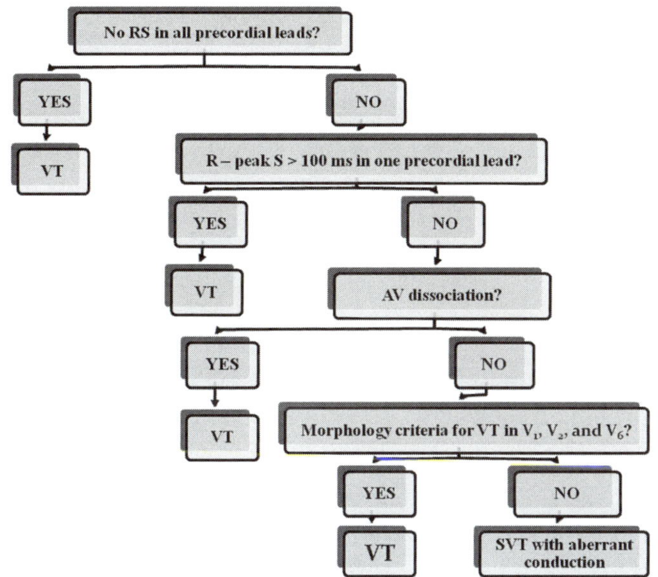

Figure 10.3. Brugada algorithm for differential diagnosis of regular wide QRS tachycardia. The morphology criteria suggestive of VT are R > R′, Rs, and qR (rsR′ and rR′ suggest SVT).

NSVT is usually asymptomatic and most often diagnosed on a routine ECG or on EST. Most of its significance depends on the presence or not of SHD.

1. NSVT in the presence of SHD

a. CAD: NSVT during the first 24 to 48 hours after acute coronary syndrome is frequent (40%–70% of patients), but it does not imply increased cardiovascular risk as it is probably related to ischemia or reperfusion; when NSVT is documented after the first week postinfarction the risk of SCD increases at least 2-fold. Later incidence usually results from scar formation, and the arrhythmia is typically monomorphic (which suggests a reentrant EP mechanism).

When symptomatic, NSVT can be treated with β-blockers, CCBs (nondihydropyridines), AADs (amiodarone, sotalol, dofetilide, procainamide), and even catheter ablation in resistant cases.

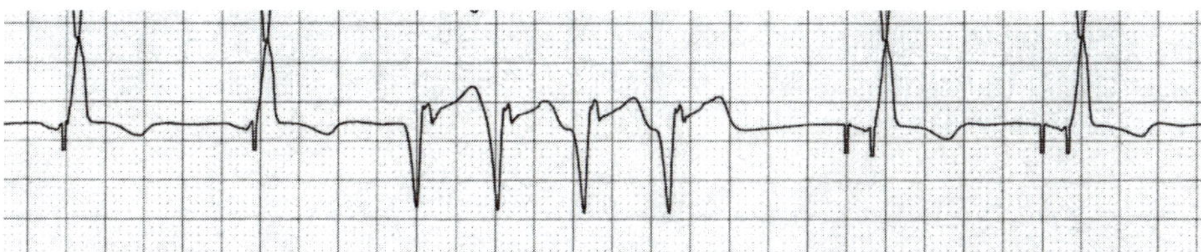

Figure 10.4. Monomorphic NSVT (4 beats at ~150 bpm).

Table 10.1. Recommendations for EPS in Patients with CAD

Class I

- For diagnostic evaluation of patients with remote myocardial infarction with symptoms suggestive of ventricular tachyarrhythmias, including palpitations, presyncope, and syncope
- To guide and assess efficacy of VT ablation
- For the diagnostic evaluation of wide QRS tachycardia of unclear mechanism

Class IIa

- For risk stratification in patients with remote myocardial infarction, NSVT, and LV EF ≤ 40%

Regardless of the presence of symptoms, the EF is the most important known marker of risk of SCD: when CAD is associated with depressed LV function (EF ≤ 40%), the risk of sudden death in the presence of ventricular arrhythmias is much higher (although this does not apply to patients with HF because their risk is already high). If the EF is combined with an EP study (Table 10.1), the information obtained helps risk-stratify the patients and determine those more likely to benefit from a prophylactic ICD implant.

In patients with EF ≤ 30% the total and arrhythmic mortalities are increased significantly and a prophylactic ICD implant is recommended (regardless of the presence or not of NSVT); no further stratification testing is necessary. EPS induces sustained monomorphic VT (SMVT) in 20% to 40% of asymptomatic patients with CAD, NSVT, and EF 30% to 40%, which confers a worse prognosis; so, if the EPS is positive, a prophylactic ICD is also recommended in this group of patients. In patients with preserved EF, NSVT can be detected in up to 5% but its presence does not appear to confer an adverse prognosis: these patients should be treated with a β-blocker as part of the standard management of patients with CAD and for the purpose of reducing the risk of SCD.

b. Hypertrophic cardiomyopathy (HCM): Asymptomatic NSVT can be detected in up to 25% of patients with HCM; its presence is associated with an increased risk for sudden death and a prophylactic ICD may be indicated (mostly in patients with prolonged episodes, repetitive or associated with symptoms). EPS may induce polymorphic VT but its significance is uncertain.

c. Nonischemic dilated cardiomyopathy (NIDCM): NSVT is very common but its presence is not associated with an increased risk of sudden death (while EF does predict mortality independently). Induction of sustained ventricular arrhythmias during EPS also has low predictive value for risk of sudden death, the test being useful mostly to diagnose bundle branch reentry (BBR) VT, to guide ablation, and to study patients with sustained palpitations, wide QRS tachycardia, presyncope, or syncope. Patients with NIDCM should receive standard therapy (β-blockers, ACEIs) and an ICD implant when eligibility criteria are met (see Chapter 12 for further information).

2. NSVT in the absence of SHD

Once adequate testing (echocardiography, EST, myocardial perfusion imaging, coronariography, etc) has excluded SHD, asymptomatic monomorphic NSVT does not require specific therapy as the prognosis is usually good (no increased risk for heart disease has been consistently found). If NSVT occurs during exercise the prognosis seems to be good as well if no ischemic ST-segment changes are induced.

Symptomatic patients can be treated initially with β-blockers and, if necessary, with amiodarone or sotalol. Episodes of repetitive monomorphic VT may cause symptoms but the risk of death is very low.

Polymorphic NSVT is an indicator of risk of sudden death and should be dealt with as soon as the diagnosis is made. Its most common etiologies are discussed following.

C. Sustained VT

VT is considered sustained if it lasts ≥ 30 seconds or if it requires termination due to hemodynamic compromise, assessed through ECG monitoring (Figure 10.5).

1. Sustained monomorphic VT

a. SMVT in the presence of SHD:

(1) General principles: IHD is by far the commonest underlying reason for SMVT, and its presence usually indicates a worse prognosis. Most episodes occur late after acute MI, sometimes many years apart; this late occurrence is most often due to scarring, but new ischemic events, severe LV dysfunction, or a ventricular aneurysm should also be ruled out. Patients with NIDCM may present with a peculiar type of SMVT (BBR VT).

(2) Mechanisms: The EP substrate for VT gradually develops in the first 2 weeks after MI and, once established, appears to remain indefinitely, even if the arrhythmia triggers are adequately controlled. These arrhythmia triggers (acute ischemia, surges in autonomic tone, onset of clinical HF) provide the link between susceptibility and spontaneous VT, which usually has a reentrant mechanism. When the VT is inducible, it signifies the presence of an anatomic VT substrate (most commonly a scar) and confirms the increased susceptibility to arrhythmia events, but its relationship with spontaneous VT is still poorly understood. Automaticity can occur in some patients with ventricular scars. A SMVT may deteriorate into VF but it is not the only clinically important mechanism of VF induction.

(3) Clinical history: The hemodynamic tolerance to a SMVT, and therefore the symptomatic status, is determined by the rate of the arrhythmia, the baseline LV function, and the development of ischemia and/or mitral regurgitation. Some patients may be mildly affected (palpitations, weakness, diaphoresis) but others may develop more significant symptoms (angina, syncope, cardiogenic shock). An association with digitalis toxicity and cocaine abuse should be excluded.

(4) Diagnosis: The ECG (Figure 10.5) in NSR may reveal clues of the underlying SHD: necrosis (CAD), LVH (HCM), epsilon waves, and/or T-wave inversion (ARVC/D), and so forth. During infarct-related SMVT it is possible to predict the endocardial LV VT exit site location based on ECG characteristics (Table 10.2).

In the lab, once a SMVT has been documented, reversible causes should always be excluded: ischemia, electrolyte disturbances, anemia, drug effects, and so forth.

Even if a SHD is known, further testing is usually necessary to exclude undiagnosed cardiomyopathies, associated anomalies, and/or progression of disease; echocardiography and nuclear imaging are most helpful in this setting. Studying new ischemic events may require coronariography.

Indications for EPS in the context of CAD were outlined in Table 10.1. The role of EPS in NIDCM is quite limited, unless a BBR VT is suspected. In patients with ARVC/D the EPS helps to reproduce the clinical VT and to guide the ablation, but the response to testing is influenced by the severity and progression of the disease. Induction of SMVT in high-risk groups—for example, SHD and syncope—may guide a decision for ICD implantation. In HCM the use of EPS remains controversial as no proven utility has been demonstrated in clinical trials. Using EPS to assess response to pharmacological therapy has largely been abandoned.

(5) Therapy: Again, all wide QRS tachycardia should be presumed to be VT until proven otherwise and, therefore, treated accordingly.

In acute episodes: If hemodynamic compromise is present, ECV, with appropriate sedation if the patient is conscious, is recommended as first-line therapy. If no

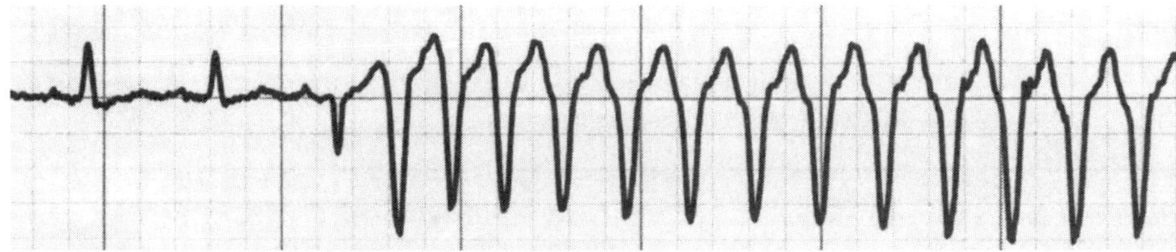

Figure 10.5. Sustained monomorphic VT.

Table 10.2. Prediction of VT Exit Site according to ECG Characteristics Reprinted from Segal OR, Chow AWC, Wong T, et al. A novel algorithm for determining endocardial VT exit site from 12-lead surface ECG characteristics in human, infarct-related ventricular tachycardia. *J Cardiovasc Electrophysiol.* 2007;18(2):161-168, by permission of John Wiley and Sons.

12-Lead ECG Patterns with Specificities or Sensitivities ≥ 70% for VT Exit Site

	BBB	Inf. leads	I	aVL	aVR	Exit	PPV %	Sensitivity %
1	RBBB	neg	pos	pos		PM/PB	100	100
2	RBBB	neg	neg	neg		PA	100	18
3	RBBB	neg	neg	pos	pos	PA	77	59
4	RBBB	pos	neg	neg		AM	77	100
5	RBBB	pos	pos	neg		AB	100	100
6	LBBB	neg	neg			SB	100	33
7	LBBB	neg	pos	pos	neg	SM	100	35
8	LBBB	neg	pos	pos	pos	SA	22	100
9	LBBB	pos	neg			SM	100	22

A A = antero-apical; AB = antero-basal; AM = mid-anterior; BBB = bundle branch block; Exit = exit site; Inf. Leads = inferior lead polarity; LBBB = left bundle branch block; neg = negative polarity; PA = posterior apex; PM = mid-posterior; pos = positive polarity; PPV = positive predictive value; RBBB = right bundle branch block; SA = antero-septal; SM = mid-septum.

pulses are palpated, the patient should be treated as if having VF. If the rate of the SMVT is too rapid, nonsynchronized shocks are preferred. IV amiodarone is reasonable in patients with recurrent VT or refractory to cardioversion (other possibly useful agents are procainamide and lidocaine). Transvenous catheter pace termination can be useful to treat SMVT that is refractory to cardioversion or is frequently recurrent despite AAD.

IV procainamide (or ajmaline) is reasonable for initial treatment of patients with stable SMVT. Other useful agents are amiodarone and sotalol. Lidocaine, though widely used in the past, seems to have only modest efficacy. If pharmacological therapy fails to control the arrhythmia, ECV with sedation should then be tried.

In patients with recurrent SMVT, or receiving multiple shocks from ICD, IV amiodarone is particularly helpful.

CCBs should not be used to terminate wide QRS tachycardia of unknown origin, especially in patients with history of myocardial dysfunction.

Chronic therapy includes AADs, ICDs, and ablation. Chronic use of AADs, other than β-blockers, in patients with ventricular arrhythmias is hindered by their limited efficacy in preventing recurrences and decreasing risk of SCD; occurrence of proarrhythmia compounds these problems, increasing the incidence of arrhythmic deaths, ischemic death, and overall mortality. AADs should not be used as primary therapy but as adjuncts to ICDs or in patients refusing or not considered candidates for an ICD. Drugs with likely better effectiveness are amiodarone, sotalol, azimilide, and dofetilide.

ICD therapy has proven superior to AADs in decreasing arrhythmic death and overall mortality. Clinical trials have included both primary (survivors of life-threatening ventricular tachyarrhythmias, including hemodynamically unstable SMVT) and secondary (prophylactic ICD implants in patients with a prior infarction, depressed LV EF, spontaneous NSVT, and inducible VT or VF; and prophylactic implants in patients with prior infarction and an LV EF < 30%–35%) indications. A more detailed discussion of ICDs and their indications is found in Chapter 12.

The recommendations for ablation of ventricular arrhythmias are found in Table 10.3. Most candidate patients are those with an ICD who have frequent episodes of monomorphic VT, for the purpose of reducing symptoms and ICD shocks.

Radiofrequency catheter ablation has a variable success rate (50%–90%) and, therefore, should not be considered an alternative to an ICD implant. The use of cool-tipped catheters, newer energy sources like cryothermy, epicardial mapping and ablation, as well as improved endocardial mapping and ablation techniques may increase the effectiveness of the procedure.

Surgical ablation as well as intraoperative catheter ablation still has a role in daily practice, though limited: candidate patients are mostly those undergoing surgery for other reasons, like aneurysmectomy or correction/completion of surgical procedures in CHD. Successful surgical or percutaneous coronary revascularization in CAD has favorable antiarrhythmic effects, with improved survival and reduction in SCD. Despite this, observational studies have suggested that SMVT in patients with previous infarction is unlikely to be affected by revascularization.

A particular group of patients with increasing diagnosis of ventricular arrhythmias is that of adult CHD: survivors of a cardiac arrest should receive an ICD implant after reversible causes have been excluded. Those with spontaneous SMVT should undergo invasive hemodynamic and EP evaluation: if catheter ablation or surgical resection of VT foci is not successful, an ICD should then be implanted.

(6) Cardiology evaluation: All patients with NSVT/SMVT should consult with a cardiologist.

b. SMVT in the Absence of SHD: In this setting, the prognosis is more benign as patients have normal hearts. The commonest arrhythmias are adenosine-sensitive VT (RVOT VT) and verapamil-sensitive VT. Propranolol-sensitive VT (with an automatic mechanism) is a less frequent type.

Table 10.3. Indications for Catheter Ablation of VT

Patients with SHD
Catheter ablation of VT is recommended

- For symptomatic SMVT, including VT terminated by an ICD, that recurs despite AADs or when AADs are not tolerated or not desired
- For control of incessant SMVT or VT storm that is not due to a transient reversible cause
- For patients with frequent PVCs, NSVTs, or VT that is presumed to cause ventricular dysfunction
- For BBR or interfascicular VT
- For recurrent sustained polymorphic VT and VF that is refractory to AAD when there is a suspected trigger that can be targeted for ablation

Catheter ablation should be considered

- In patients who have one or more episodes of SMVT despite therapy with one or more class I or III AAD
- In patients with recurrent SMVT due to prior infarction who have LV EF > 30% and expectation for 1 year of survival, and is an acceptable alternative to amiodarone therapy
- In patients with hemodynamically tolerated SMVT due to prior infarction who have reasonably preserved LV EF (> 35%) even if they have not failed AAD

Patients without SHD
Ablation is recommended for patients with idiopathic VT

- For monomorphic VT that is causing severe symptoms
- For monomorphic VT when AADs are not effective, not tolerated, or not desired
- For recurrent sustained polymorphic VT and VF (electrical storm) that is refractory to AAD when there is a suspected trigger that can be targeted for ablation

VT Catheter Ablation Is Contraindicated

- In the presence of a mobile ventricular thrombus (epicardial ablation may be considered)
- For asymptomatic PVCs and/or NSVT that are not suspected of causing or contributing to ventricular dysfunction
- For VT due to transient, reversible causes, such as acute ischemia, hyperkalemia, or drug-induced TdP

(1) RVOT VT:

- **General principles:** RVOT VT is the most common type of idiopathic VT (70%–80% of all cases). It usually has a benign course as 60% to 80% of those affected have structurally normal hearts. Two phenotypes are recognized: nonsustained, repetitive monomorphic VT (60%–92% of cases) and paroxysmal, exercise-induced VT. The diagnostic of ARVC/D has to be ruled out, especially if the baseline ECG is abnormal.
- **Mechanisms:** Triggered activity mediated by catecholamine-induced delayed afterdepolarizations (phase 4 of the AP).
- **Clinical history:** Symptoms typically start between 20 and 50 years of age, most often in women. Palpitations with exercise are the most common presentation but presyncope/syncope may occur. RVOT VT is induced by EST in 25% to 50% of cases, during the exercise or in the recovery period; if repetitive monomorphic VT is present, it is often suppressed during exercise. Other known triggers are caffeine, stress, and premenstrual states.
- **Physical examination:** Usually normal if patient is in NSR.
- **Diagnosis:** ECG is usually normal when in NSR; the ECG during the arrhythmia shows monomorphic VT with LBBB morphology and inferior axis, indicating origin in the RV outflow tract; precordial transition in V_3 to V_4 (this finding suggests a septal origin; transition beyond V_4 suggests a free-wall site of origin); typical large monophasic R waves in inferior leads, suggesting a septal origin (notching suggests a free-wall origin); CL oscillations are common.

  Analysis of lead I further helps to locate the site of origin of the arrhythmia: free-wall and septal rightward and posterior locations have a positive QRS polarity (R or r waves), leftward and anterior sites along the septum and the free wall have a negative polarity (s or QS pattern), and intermediate sites have a biphasic or multiphasic QRS morphology (qr/rs pattern) and/or an isoelectric segment preceding an r wave.

  In 10% of cases the arrhythmia arises from the LVOT; the VT then has RBBB morphology and there is early precordial transition (RS ratio ≥ 1 in V_1–V_2).

  Laboratory testing includes routine blood tests, which are often normal.

  The echocardiogram is usually normal but the cardiac MRI could show abnormalities somewhat similar to those found in ARVC/D.

  EST serves to correlate symptoms with the physical stress and to assess drug efficacy.

  EPS is carried out to establish the precise diagnosis and to guide the ablation procedure. Typical EP characteristics are summarized in Table 10.4.

  Most arrhythmia foci are localized in the anterosuperior aspect of the interventricular septum, under the pulmonary valve, or in the free wall.

Table 10.4. EP Characteristics of RVOT VT

- Rarely inducible in sedated patients
- Poor day-to-day reproducibility due to extreme sensitivity to autonomic influences
- If isoproterenol is utilized in an attempt to induce the arrhythmia, abrupt withdrawal may allow triggering the arrhythmia during the washout phase
- It may be initiated by using atropine, aminophylline, or burst atrial or ventricular pacing
- The VT cannot be entrained
- Termination with adenosine, Valsalva maneuvers, carotid sinus massage, edrophonium, verapamil, and β-blockers

- **Therapy:** Medical management of acute episodes include maneuvers that increase the vagal tone or using drugs such as IV amiodarone, β-blockers, procainamide, adenosine, or verapamil. Long-term management includes β-blockers, verapamil, and AADs (III, Ia, and Ic). Nonpharmacological therapy includes the following: In very symptomatic patients (e.g., syncope), with very rapid VT, poorly responsive to medical therapy or drug-intolerant, catheter ablation is indicated (Table 10.3), with 90% to 95% success rate. The use of tridimensional activation mapping helps to improve the spatial resolution of pacemapping. Cryoablation, thanks to the firm adherence of the catheter to the myocardial tissue throughout the whole ablation time, may help improve the success rate. In experienced centers major complications are ≤ 1% and the risk of recurrence is ~5%.
- **Cardiology evaluation:** Suggested in symptomatic cases, when AAD or catheter ablation is contemplated.

(2) Verapamil-sensitive VT:

- **General principles:** The most common form of idiopathic left VT; usually diagnosed at ages 15 to 40; males make up 60% to 80% of those affected. Verapamil-sensitive VT can rarely be incessant, leading to tachycardia-induced cardiomyopathy. SCD is usually not associated to it.
- **Mechanisms:** Verapamil-sensitive VT originates from the region of the left posterior fascicle (exit site), near the inferoposterior LV septum. It has a reentrant mechanism: the proposed circuit has an antegrade limb including abnormal Purkinje tissue (characterized by slow decremental conduction, sensitivity to verapamil, giving rise to diastolic potentials along the midseptum) and a retrograde limb formed by Purkinje tissue from the left posterior fascicle (or contiguous to it), which gives rise to Purkinje potentials (this circuit explains why, with successful ablation, there is a progressive increase

in diastolic potential–Purkinje potential interval whereas the Purkinje potential–ventricular activation interval remains constant).

- **Clinical history:** Most of the episodes occur at rest. Palpitations and presyncope are frequent symptoms but cardiac arrest is very rare.
- **Diagnosis:** ECG reveals RBBB morphology, with left (superior) axis deviation (90%–95% of cases); QRS relatively narrow (120–140 milliseconds). In 5% to 10% of cases, there is RBBB morphology with right inferior axis deviation, indicating exit site in the left anterior fascicle, near the anterosuperior LV septum.
  Echocardiography is usually normal.
  EPS may be needed to establish the diagnosis (Table 10.5).
  During EPS, differentiation from an interfascicular VT (Table 10.6) is of particular importance as the ablation procedure differs significantly.
- **Therapy:** In medical management, acute episodes are treated with IV verapamil; its long-term use seems also to be effective in preventing recurrences. Other drugs that can be tried are β-blockers, CCBs, sotalol, amiodarone, and flecainide. Nonpharmacological therapy for those patients with presyncope/syncope, who are drug-intolerant/-resistant or with recurrence despite adequate medication, should include catheter ablation, as success rate is > 90%.
- **Cardiology evaluation:** Verapamil-sensitive VT is better managed by specialized physicians.

2. Polymorphic VT

Polymorphic VT is a VT characterized by variable HR and irregular rhythm; the QRS complexes are wide, with a phasic variation. It may occur in the presence of normal QT interval (acute infarction, CPVT, BrS, vasospastic angina) or with long QT (TdP). The involved EP mechanism is reentry or triggered activity.

Table 10.5. EP Characteristics of Verapamil-sensitive VT

Not terminated by Valsalva or adenosine
If isoproterenol is required to initiate it, it then becomes adenosine-sensitive (all reentrant VT with SHD that require isoproterenol for facilitation are uniformly insensitive to adenosine)
Initiated by atrial pacing
IV verapamil significantly prolongs the VT CL
The local activation order during VT is always diastolic potential–Purkinje potential–ventricle (during NSR, Purkinje potential activates before diastolic potential)

Table 10.6. EP Characteristics of Interfascicular VT

- RBBB during tachycardia, with left anterior fascicular block or right axis deviation when conduction is antegradely over the left anterior fascicle and retrogradely over the left posterior fascicle (a reverse activation sequence gives a left axis deviation)
- Reverse activation sequence of His and left bundle branch potentials during tachycardia (left bundle branch potential before the His)
- HV shorter during VT than during sinus rhythm, by > 40 ms
- Spontaneous oscillation in CL of the tachycardia due to a change in the LBB-LBB intervals that precede and drive the CL
- Termination of tachycardia with ventricular extrastimuli or radiofrequency that produces block in either of the 2 fascicles

Vasospastic angina as a cause of polymorphic VT has been recently recognized; it is often associated with AF and it has a high incidence of SCD. The VT is preceded by ST-segment elevation and initiated by R-on-T PVCs, long-short sequence of PVCs, or TWA. Treatment is with CCBs and ICD implant.

Polymorphic VT may be sustained or self-terminating, alternating with NSR. In patients with hemodynamic compromise, prompt ECV is recommended. In the context of acute ischemia, IV β-blockers and amiodarone are useful (lidocaine may be tried as well); intra-aortic balloon pump and urgent revascularization are often indicated.

a. Catecholaminergic Polymorphic VT

(1). General principles: CPVT is an inherited arrhythmia disorder characterized by adrenergically-mediated polymorphic ventricular tachyarrhythmias. The majority of events occur during childhood; by age 20, > 60% of patients have already suffered syncope or cardiac arrest. If left untreated, mortality is 30% to 50% by age 40.

(2) Mechanisms: CPVT is a genetic disorder: it is transmitted in an autosomal dominant form associated with mutations in the gene encoding the cardiac ryanodine receptor (*RyR2*), and a recessive form associated with mutations in the gene encoding the cardiac isoform of calsequestrin (*CASQ2*). These mutations allow for delayed afterdepolarizations and triggered activity to originate the typical morphology of CPVT.

(3) Clinical history: There is a history of exercise or emotion-related palpitations and dizziness in some patients. Syncope during physical activity or acute emotion is present in > 60% cases (which makes a good case for differential diagnosis with LQTS). Typically the arrhythmias are reproducibly inducible with EST, with increasing workloads leading to increased severity (i.e., isolated PVCs lead to NSVT, which leads to sustained VT, self-terminating or leading to VF). Supraventricular arrhythmias may also be induced by exercise.

(4) Diagnostics: The resting ECG is typically normal; some patients may have sinus bradycardia with prominent U waves. The tachycardia has typical features: a bidirectional VT with beat-to-beat 180° rotation of the QRS complex. Irregular polymorphic VT is also seen. Imaging shows a normal heart in most patients.

(5) Therapy: β-blockers (e.g., nadolol) are effective in 60% of patients, preventing recurrence of the arrhythmias. They should be used at maximum tolerated doses and repeat EST is necessary to assess their efficacy. If ventricular arrhythmias are still present during EST or if syncope recurs in patients compliant with full dose of β-blockers, an ICD implant should be considered. High doses of β-blockers should be maintained after ICD implantation because ICD shocks cause an hyperadrenergic state, which predisposes to further VT. Sympathetic denervation may be of utility in patients with recurrent arrhythmias or frequent ICD discharges despite high-dose β-blockade.

(6) Cardiology evaluation: CPVT is better managed by cardiologists.

b. Torsade des Pointes

(1) General principles: TdP is characterized by twisting of the peaks of the QRS complexes around the isoelectric line during the episode of VT. It occurs in the setting of long QT interval and in patients with advanced conduction system disease leading to heart block. More frequent in females (probably because they have longer QT interval), it is usually recurring and short lasting, but it may turn into sustained VT/VF. TdP is life-threatening and could present as SCD in patients with structurally normal hearts.

(2) Mechanisms: Triggered by early afterdepolarizations during phase 3 of the AP.

(3) Clinical history: TdP should be suspected in any patient with pause-dependent VT, though in congenital LQTS in children it is not pause-dependent (see Chapter 6). Patients often are on medications that prolong the QT interval (AADs, antibiotics, antihistaminergics) or suffer from electrolyte disorders (e.g., hypokalemia, hypomagnesemia).

(4) Diagnosis: ECG: Ventricular bigeminy in a patient with long QT interval is a sign of impending TdP. In the context of acquired bradyarrhythmias a prolonged QT interval (≥ 510 milliseconds), LQT2-like morphology (notched T waves), and

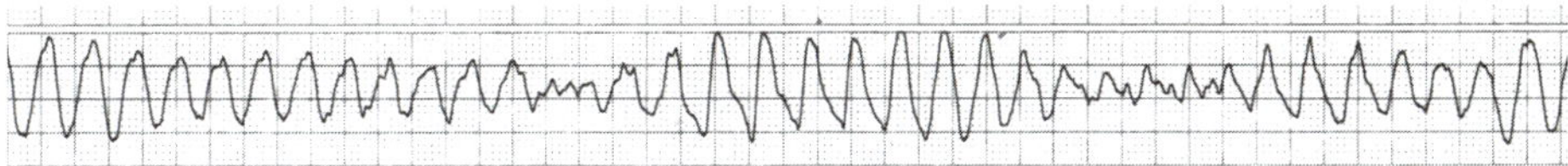

Figure 10.6. TdP in a patient with acquired LQTS.

$T_{peak} - T_{end} \geq 85$ milliseconds also correlate with increased risk for TdP. The arrhythmia has peculiar features (Figure 10.6): early afterdepolarizations may appear as pathological U waves; typically initiated following "short-long-short" coupling intervals (PVC-compensatory pause-PVC initiating TdP) (there is also a short coupled variant initiated by normal-short coupling); HR 200 to 250 bpm; irregular rhythm; wide QRS complex with continuously changing morphology, alternating from positive to negative; TWA may be seen before it; and prolonged QT interval.

In laboratory testing, electrolyte levels should be checked out (K^+ is usually low); ischemia must be excluded. Echocardiography is the preferred imaging technique to rule out SHD.

(5) Therapy: Discontinue offending agent (diuretics are common etiology, independent of the serum electrolyte levels); treat electrolyte abnormalities (keep $K^+ > 4.5$ mmol/L); suppress early afterdepolarizations with intravenous magnesium: 1 to 2 g over 30 to 60 seconds, followed up by infusion of 3 to 10 mg/min. Increase HR with isoproterenol (> 90 bpm) or temporary pacing (90–110 bpm until QT normalizes). Isoproterenol is first option in pause-dependent TdP but is contraindicated in cases of congenital LQTS (together with all other β-adrenergic agents). Pacing is the first option when TdP is due to heart block or symptomatic bradycardia (most of these patients will end up with permanent pacing).

ECV as needed.

Intravenous lidocaine and oral mexiletine may be tried in patients with LQT3 and TdP. Patients with LQTS need specific long-term management (see Chapter 6).

(6) Cardiology evaluation: Usually not necessary when TdP is nonsustained and due to transitory causes (e.g., diuretic use). In patients with LQTS or indication for permanent pacing, specialized consult is recommended.

3. Incessant VT

a. General principles: Incessant VT, or VT storm, is 3 or more separate episodes of sustained VT within 24 hours, each requiring termination by an intervention, or a VT present > 50% of a 24-hour period. Most common causes are acute myocardial ischemia; exacerbation of HF; metabolic abnormalities such as hypokalemia, hypomagnesemia, and hyperthyroidism induced by amiodarone; change in prescribed medications, most importantly AADs; proarrhythmia effect; and noncompliance with prescribed medications.

b. Therapy: The management of VT storm depends chiefly on its ECG presentation, which correlates well with its cause. Incessant polymorphic VT tends to be ischemic in origin; revascularization and β-blockade should be pursued aggressively, oftentimes aided by IV amiodarone or procainamide. With incessant monomorphic VT, the most likely mechanism is reentry around a scar. IV amiodarone or procainamide tend to control it, and VT ablation should be considered. The combination of IV amiodarone and IV β-blockers often allows for good control. For patients not responding well to these measures, overdrive pacing or general anaesthesia can be of help; sympathetic denervation has also been tried with success.

V. BIDIRECTIONAL VT

A. General Principles

Bidirectional VT is an uncommon type of VT, where QRS complexes have an alternating morphology. It tends to be self-terminating, lasting from a few seconds to a few minutes.

Its most common cause is digitalis toxicity but may also be seen in severe SHD, CPVT, familial hypokalemic periodic paralysis (Andersen-Tawil syndrome), and herbal aconite poisoning, as well as in normal hearts.

B. Mechanisms

Bidirectional VT seems to be originated by triggered activity or, less commonly, enhanced automaticity. The exact site of origin is not clear; case reports have demonstrated bidirectional VT due to ventricular bigeminy in the presence of junctional tachycardia, and also due to 2 separate ventricular foci. There have also been reports of bidirectional tachycardia with a supraventricular origin, due to an SVT with associated trifascicular disease (with permanent aberrant conduction in the right bundle branch alternating with aberrant conduction in the 2 fascicles of the left bundle branch).

C. Clinical History

An association with digitalis use has been found in 82% of cases. A history of CAD or severe SHD is also common.

D. Diagnostics

1. ECG

ECG shows beat-to-beat alternating morphology and axis of QRS complexes; ventricular rate 140 to 180 bpm; QRS usually wide, with RBBB morphology; and rhythm may be regular or irregular.

2. EP Study

Not usually needed.

E. Therapy

Treatment of bidirectional VT includes correction of reversible factors like decompensated HF and electrolyte abnormalities. When present in patients taking digitalis, it is highly suggestive of toxicity and antidigoxin antibody fragments are usually needed as therapy (ECV should be avoided). As a temporary measure some AADs may be used: IV lidocaine is first choice, with quinidine and flecainide as second options. Slow bidirectional VT with hemodynamic compromise may be treated with overdrive atrial pacing.

VI. BUNDLE BRANCH REENTRY VT

A. General Principles

BBR VT is one of the 3 tachyarrhythmias due to reentry within the His-Purkinje system, the others being the isolated, repetitive complexes commonly seen during ventricular EPS (V_3 phenomenon) and the interfascicular VT. BBR VT should be suspected in patients with VT and evidence of His-Purkinje system disease in the ECG (long PR, wide QRS, partial or complete LBBB). It is more frequent in cases of NIDCM (40% of VT in this setting is attributed to it); it may also be diagnosed in the first 1 to 2 weeks after aortic or mitral valve surgery. BBR VT is associated with syncope/SCD and its diagnosis should be followed up by aggressive therapy.

B. Mechanisms

BBR VT is due to reentry among the bundle of His and the bundle branches; the circuit is usually initiated by PVCs, which conduct retrogradely up the left bundle branch (the impulse gets blocked in the right branch due to a longer RP) to the bundle of His, leading to its activation. The impulse then may conduct antegrade down the right bundle branch, slowly so that allows recovery of conduction through the left branch and its retrograde depolarization, which closes the circuit and allows tachycardia to be established.

C. Clinical History

Most affected patients have severe heart disease, either ischemic or nonischemic. A history of HF is usually elicited. Presyncope, syncope, and aborted SCD are common presenting symptoms.

D. Diagnostics

1. ECG

ECG shows fast-rate VT, usually > 200 bpm; typical LBBB morphology (the intrinsicoid deflexion could be narrow, confusing it with SVT); leftward axis deviation (this indicates left anterior fascicular block, excluding this fascicle from being part of the circuit).

Most patients have a similar ECG during sinus rhythm because of the underlying prolongation of the HV interval. A reverse circuit is less frequent, with RBBB morphology and right/normal axis.

2. Laboratory testing

Addressed to the underlying pathology.

3. Imaging

According to needs: echocardiography, coronariography, and so forth.

4. EP Study

Needed to confirm the diagnosis and to guide the ablation procedure. The EP features are found in Table 10.7.

E. Therapy

1. Medical management

AADs are usually not helpful. ECV may be necessary in symptomatic patients.

Table 10.7. EP Characteristics of BBR VT

- Increased HV interval, more during tachycardia than during sinus rhythm
- Increased QRS duration
- Tachycardia induction depending on a critical delay in the His-Purkinje system
- Tachycardia terminates with block in the His-Purkinje
- A His complex precedes each QRS; it is activated retrogradely, simultaneously with the proximal part of the bundle branch (which has antegrade activation)
- Stable HV interval
- Variations in VV interval are preceded by changes in HH interval
- Postpacing interval–tachycardia CL ≤ 30 ms with entrainment from apex of RV
- QRS morphology during atrial entrainment is identical to that during tachycardia (concealed fusion); during ventricular entrainment there is manifest fusion

2. Nonpharmacological therapy

Catheter ablation of the right bundle branch, leading to complete RBBB, will eliminate the tachycardia and is the therapy of choice for all patients. Infranodal conduction may worsen significantly, making permanent pacing necessary.

Due to the high risk of SCD an ICD implant should be considered (mostly if there is inducible VT during EPS postablation); nevertheless, most patients with BBR VT already meet criteria for ICD therapy for primary prevention of SCD.

F. Cardiology Evaluation

All patients with suspected BBR VT should be evaluated and managed by cardiologists.

VII. VENTRICULAR FLUTTER

A. General Principles

Ventricular flutter is a regular (CL variability ≤ 30 milliseconds) VT at approximately 300 bpm (CL 200 milliseconds) with a monomorphic appearance and no isoelectric interval between successive QRS complexes (the term *monomorphic VT with indeterminate QRS morphology* is currently preferred). Ventricular flutter is a possible transition stage between VT and VF and requires immediate management.

B. Mechanisms

Its EP mechanism is reentry.

C. Diagnostics

The ECG shows regular, large oscillations at a rate of 250 to 300/min, resembling a sine wave or a helix.

D. Therapy

Ventricular flutter should be treated with immediate defibrillation or unsynchronized cardioversion.

VIII. VENTRICULAR FIBRILLATION

A. General Principles

VF is a rapid, usually > 300 bpm (CL 200 milliseconds), grossly irregular ventricular rhythm with marked variability in QRS CL, morphology, and amplitude. In VF there is no organized cardiac electrical activity and cardiac output is practically nonexistent.

It leads to asystole if left untreated and is the most recorded rhythm at the time of sudden cardiac arrest. Spontaneous VF is rare in structurally normal hearts, but events like ischemia may trigger it in hearts with predisposing conditions.

B. Mechanisms

VF does not have a uniform origin, and there are different proposed mechanisms for different anatomic substrates:

- **Heart failure:** In these patients, macroreentrant circuits or rotors lead to VF. These rotors may be either stationary ("mother rotor") with daughter wavelets, or drifting, wandering wavelets. It is possible that VF in humans is not usually maintained by a large number of simultaneous rotors but only a few ones, and sometimes only one at any given time.
- **Ischemia:** The proposed mechanism is an initial focus transitioning to a reentry mechanism.
- **Normal hearts:** VF may be induced with a critically timed electrical stimulus during the early phase of repolarization (R-on-T phenomenon), immediately after the RP (near the peak of the T wave). A critically timed mechanical trauma to the chest (*commotio cordis*) may also lead to VF.
- VF may also be seen as degeneration of all types of sustained VT, preexcited SVTs, myocarditis, cardiac tamponade, pulmonary embolism, ion channelopathies, electrolyte disorders, and so forth.

C. Clinical History

Some patients may have chest pain, palpitations, dyspnea, or syncope before VF develops. In monitored patients, PVCs or VT may be initially documented. VF often leads to SCD, without prodromes.

D. Diagnosis

In an ECG documentation of wide, very irregular QRS complexes, paired with cardiac arrest, is confirmatory (Figure 10.7).

E. Therapy

Survival of affected patients depends on appropriate life-support measures. Once VF is documented, electrical external defibrillation should be performed at once (see Chapter 13) as it remains its most successful treatment. In patients with recurrent VF, electrolyte disorders should be ruled out. Amiodarone infusion can be of help. If ischemia is suspected as the cause, emergent revascularization must be attempted.

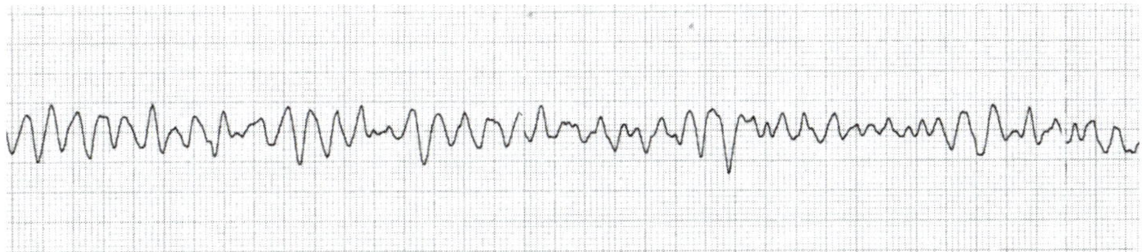

Figure 10.7. Ventricular fibrillation.

In hypothermic patients, reanimation should be continued until core temperature is increased.

F. Cardiology Evaluation

Affected patients need admission to an intensive care unit, under management from specialized personnel.

SUGGESTED READINGS

Katritsis DG, Camm AJ. Nonsustained ventricular tachycardia: where do we stand? *Eur Heart J.* 2004;25:1093-1099.

Zipes DP, Camm AJ, Borggrefe M, et al. ACC/AHA/ESC 2006 guidelines for management of patients with ventricular arrhythmias and the prevention of sudden cardiac death. *J Am Coll Cardiol.* 2006;48:247-346.

PART 4

Pacemakers, Defibrillators, and Techniques

Chapter 11

Permanent Pacing

Miguel A. Barrero Garcia and Bernard Thibault

I. GENERAL PRINCIPLES

In selected cases of bradycardia and AV conduction disorders (see Chapters 3 and 4) the implantation of a permanent pacemaker serves to relieve symptoms, improve quality of life, and decrease morbidity/mortality. Artificial pacemakers have evolved significantly over the years, and their implantation is no longer considered a major surgical procedure. Extreme care should be exercised in determining who may benefit from this therapy, selecting the correct pacing modality, and ensuring appropriate follow-up, which includes regular monitoring for possible complications and troubleshooting of pacing system malfunctions.

The following are basic concepts in pacing:

- **Voltage (V):** Force that causes electrons to move across a circuit; often referred to as amplitude.
- **Current (I):** Flow of electrons in a circuit, measured in milliamps (mA).
- **Impedance (R):** Force that opposes current flow, measured in ohms and often referred to as resistance. Reducing electrode size increases R, decreases current drain, and increases longevity of the pacemaker.
- **Ohm's law:** Describes the relationship between V, I, and R (V = I x R). Assuming a constant V, normal R results in normal I, low R results in high I, and high R leads to low I.
- **Stimulation threshold:** The minimum electrical stimulus output needed to consistently capture the heart outside of its refractory period (RP). Its measurement ensures proper lead placement and helps determine the safety pacing margin. The stimulation threshold is affected by lead maturation, lead technology, medications, and lead location.
- **Pulse amplitude:** The voltage delivered by the electrical pacing stimulus.
- **Pulse width (PW):** Duration of the artificial electrical impulse.
- **Output:** V x PW.
- **Strength-duration curve:** Represents the relationship between pulse voltage and PW (Figure 11.1), as described by the concepts of rheobase and chronaxie. Rheobase is the least voltage that will stimulate the myocardium at any given PW. Chronaxie is the threshold PW at twice rheobase voltage, thus representing the energy-efficient safety margin: when the PW is greater than chronaxie there is a waste of stimulation energy without significant increase in safety margin. If the PW is less than chronaxie then there is a steep increase in voltage and stimulation energy. As can be inferred from the curve in Figure 11.1, small changes in PW are associated with significant changes in voltage at short PWs (usually < 0.25 milliseconds) but only with small changes at longer PW (> 1 milliseconds).
- **Energy:** The pacing stimulus energy $E = \frac{v^2 \text{ x } T}{R}$, where T represents pulse duration. From this formula: 2 x V = 4 x E but 2 x PW = 2 x E; therefore,

increasing voltage is more effective in augmenting the stimulation energy than increasing PW.

- **Sensing:** The ability of a pacemaker to detect when an intrinsic depolarization is occurring. Programming a high setting decreases sensitivity and fewer signals are sensed; lowering the setting increases the sensitivity and thus more signals are sensed. Accurate sensing requires filtering out extraneous signals like T waves, far-field potentials, and myopotentials. Sensing may be affected by the EP properties of the myocardium, the characteristics of the electrode, lead placement, sensing amplifiers of the pacemaker, lead polarity and integrity, and electromagnetic interference.
- **Slew rate (dV/dt):** A measure of frequency content in the intracardiac electrogram (EGM). If the slew rate is low the signal may not be sensed even if the peak-to-peak amplitude of the EGM is large (the higher the slew rate, the higher the frequency content). Maximum density of frequencies for R waves is 10 to 30 Hz.

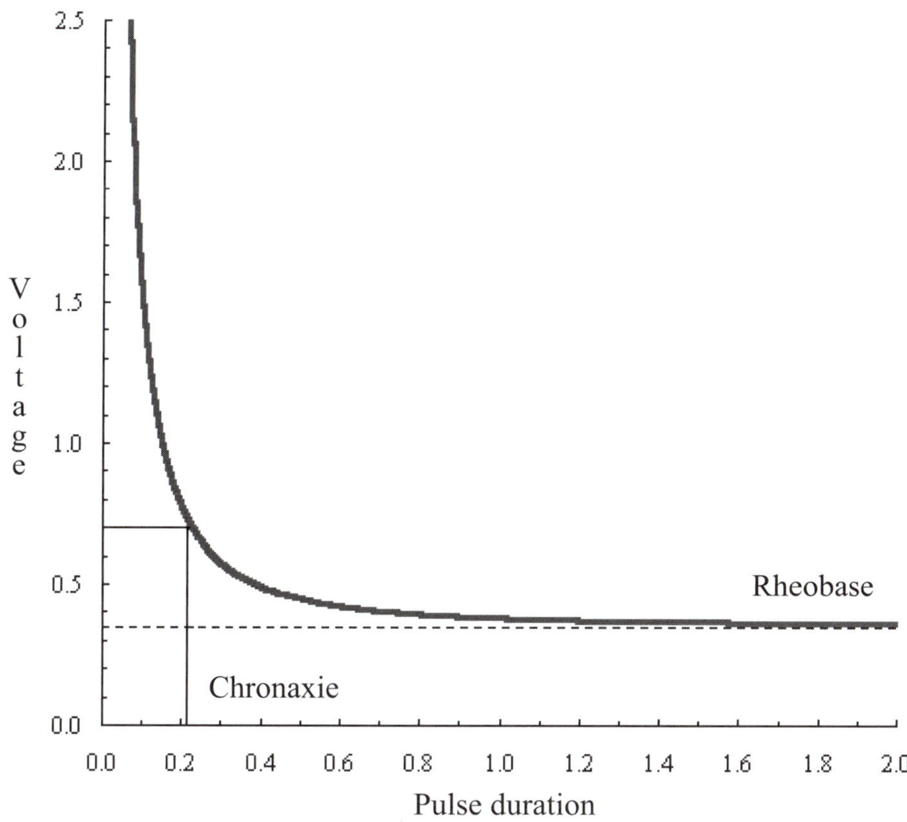

Figure 11.1. Strength-duration curve.

II. PACING CODES AND MODES

The revised NBG code is a standardized classification of pacemakers to facilitate their use and understanding (Table 11.1). Based on this code, a DDD pacing mode means the pacemaker is capable of dual (atrial and ventricular) sensing and pacing, as well as dual response to sensing; for example, a sensed atrial event will inhibit the atrial output but will trigger a ventricular stimulus (a programmable delay between a sensed P wave, and a paced ventricular depolarization allows inhibition of the ventricular stimulus if a native ventricular signal is sensed). The fourth position indicates whether a pacemaker has a sensor incorporated to adjust the programmed pacing rate to the patient activity. Because multisite pacing is not commonly used the fifth position is usually omitted.

The most commonly used pacing modes are VVI, DDD, and AAI (Tables 11.2 and 11.3), with or without rate responsiveness.

Table 11.1. The Revised NASPE/BPEG (NBG) Code

Position	I	II	III	IV	V
Category	Chamber(s) paced	Chamber(s) sensed	Response to sensing	Rate modulation	Multisite pacing
Letters Used	O, A, V, D	O, A, V, D	O, T, I, D	O, R	O, A, V, D
Manufacturer's Designation	S (A or V)	S (A or V)	—		

O = none, A = atrial, V = ventricular, D = dual, T = triggered, I = inhibited, R = rate modulation, S = single.

Table 11.2. Pacing Modes in Single-Chamber Pacemakers

Pacing Mode	AOO	AAI	VOO	VVI
Mode of Sensing	No sensing	A	No sensing	V
Mode of Pacing	A	A	V	V
Response to Presence or Absence of Intrinsic Beats	Asynchronous pacing	Sensed intrinsic P inhibits pacing	Asynchronous pacing	Sensed intrinsic QRS inhibits pacing
Ideal Patient	Temporary use	SND with intact AVN function	Temporary use	Permanent AF with slow ventricular response

Table 11.3. Pacing Modes in Dual-Chamber Pacemakers

Pacing Mode	DOO	VDD	DDI	DDD
Mode of Sensing	No sensing	A and V	A and V	A and V
Mode of Pacing	A and V	V	A and V	A and V
Response to Presence or Absence of Intrinsic Beats	Asynchronous pacing	Intrinsic QRS inhibits V pacing, intrinsic P can trigger V pacing	No tracking of P waves	Intrinsic P and QRS can inhibit pacing, intrinsic P can trigger V pacing
Ideal Patient	Temporary use	Normal sinus node function with AVB	Atrial tachyarrhythmia	AVB with normal sinus node function

Single-chamber systems are easier to implant (only one lead is required), but in the case of atrial-based systems there should be no AV conduction disturbances (many physicians would prefer to implant a dual-chamber pacemaker in a patient with sinus node dysfunction (SND) and normal AV conduction, due to a small but definitive risk of AV conduction abnormalities over time) and, in the case of ventricular-based systems, no AV synchrony can be maintained, and the patient could develop a pacemaker syndrome.

Asynchronous pacing, where the pacemaker paces at a fixed rate and sensing is suspended, is used mainly on a temporary basis such as in a pacing-dependent patient undergoing a surgical procedure (to avoid inhibition of the pacemaker by electrocautery).

III. PRINCIPLES OF PACEMAKER IMPLANTATION AND FOLLOW-UP

A pacing system includes the lead(s) and the generator. Because of the risk of bleeding, patients on oral anticoagulation should have their medication stopped at least 3 days before the procedure; depending on the anticoagulation indication, tapering with standard IV heparin is carried out. Heparin should be stopped about 4 hours before the procedure, and restarted 8 to 12 hours after; warfarin can be reinitiated on the same day. Prophylactic antibiotic coverage is common practice.

Intracardiac access is obtained from the subclavian or cephalic veins, the latter less commonly used as it has a high percentage of anatomic malformations. In most instances, a left-sided access is preferred as most patients are right-handed and because there is a less acute angle between the left subclavian and the innominate veins. Intravenous access ipsilateral to the side of the implant is advisable in case the use of contrast agents is necessary. The preferred site of pacemaker generator placement is the subclavian area, for several reasons: enough space to create the subcutaneous pocket, the subclavian vein is located close by, and relative ease of access for IV electrode implantation (an abdominal implant with epicardial electrodes is preferred in pediatric patients).

Once the leads are in place the operator uses available means, fluoroscopic and electric, to ensure an adequate position has been attained: apex of RV for the ventricular electrode and high RA (appendage) for the atrial electrode. Once the radiological position of the electrode(s) has been verified, adequate lead measurements and sensing/pacing thresholds are then validated (Table 11.4, Figures 11.2 and 11.3).

Table 11.5 shows the ECG characteristics of several electrode positions, while Table 11.6 shows some characteristics of coronary sinus pacing, one of the commonest malpositions.

The risk associated with the surgical procedure is very low and should always be adequately explained to the patient; obtaining a signed informed consent is standard practice. Infection, pneumothorax, and lead displacement all have a risk of ~1% (though this figure very much depends on the center); perforation of the heart with pericardial effusion/tamponade occurs in < 1% of patients. Bleeding risk is higher for patients on oral anticoagulation. The risk of death is very low.

Most patients can be safely discharged after a 24-hour monitoring period, once a chest x-ray shows a stable lead positioning and no immediate complications are present. Repeat testing of sensing/pacing thresholds the morning after the procedure is recommended.

Discharged patients should be reviewed in dedicated clinics at a minimum of 4 to 6 weeks postimplant to exclude surgery-related complications and to verify

Table 11.4. Recommended Thresholds for Adequate Sensing and Pacing

Parameter	Atrial Lead	Ventricular Lead
Pacing	< 1.5 V	< 1 V
Sensing	> 1.5 mV	> 6 mV
Lead impedance	400–1200 Ω	400–1200 Ω
Slew rate	> 0.2 V/s	> 0.5 V/s

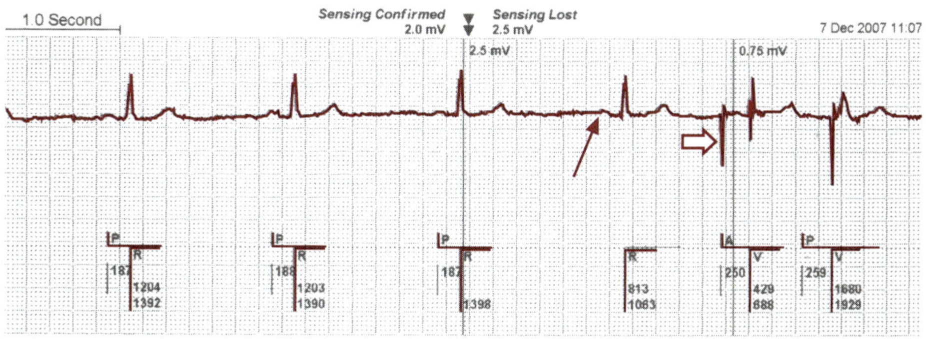

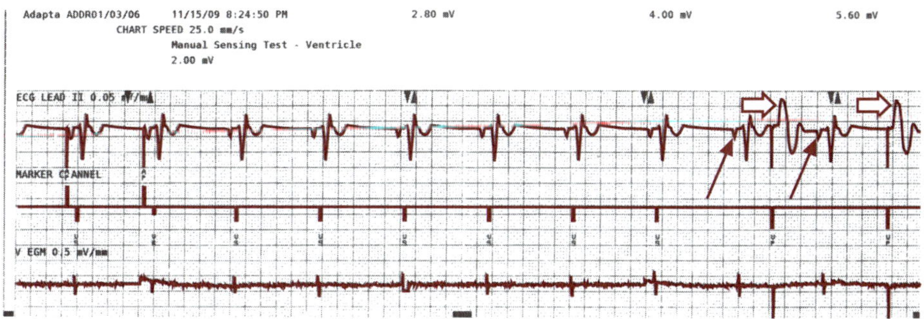

Figure 11.2 (a) Testing for atrial sensing. Adequate sensing of native events (P) with sensibility set at 2 mV; when sensibility is decreased to 2.5 mV a native event (arrow) is not detected (also notice no sensed P in the marker channel) and a paced stimulus (block arrow) ensues. (b) Testing for ventricular sensing. When ventricular sensitivity is decreased to 4 mV, native events (arrows) are not sensed and ventricular pacing starts (block arrows).

Table 11.5. ECG Markers of Electrode Position

Electrode Position	QRS Morphology	QRS Axis
RV apex	LBBB	Superior
RV inflow tract	LBBB	Normal
RV outflow tract	LBBB	Inferior or normal
Mid or high LV	RBBB	Inferior or normal
Inferior LV	RBBB	Superior
Coronary sinus	RBBB	Inferior
Cardiac veins	RBBB	Superior

adequate functioning of the pacemaker. Several steps should be taken to ensure a proper evaluation (Table 11.7). Periodic follow-up reviews should then be scheduled every 6 to 12 months for at least 5 years postimplant; older implants should be

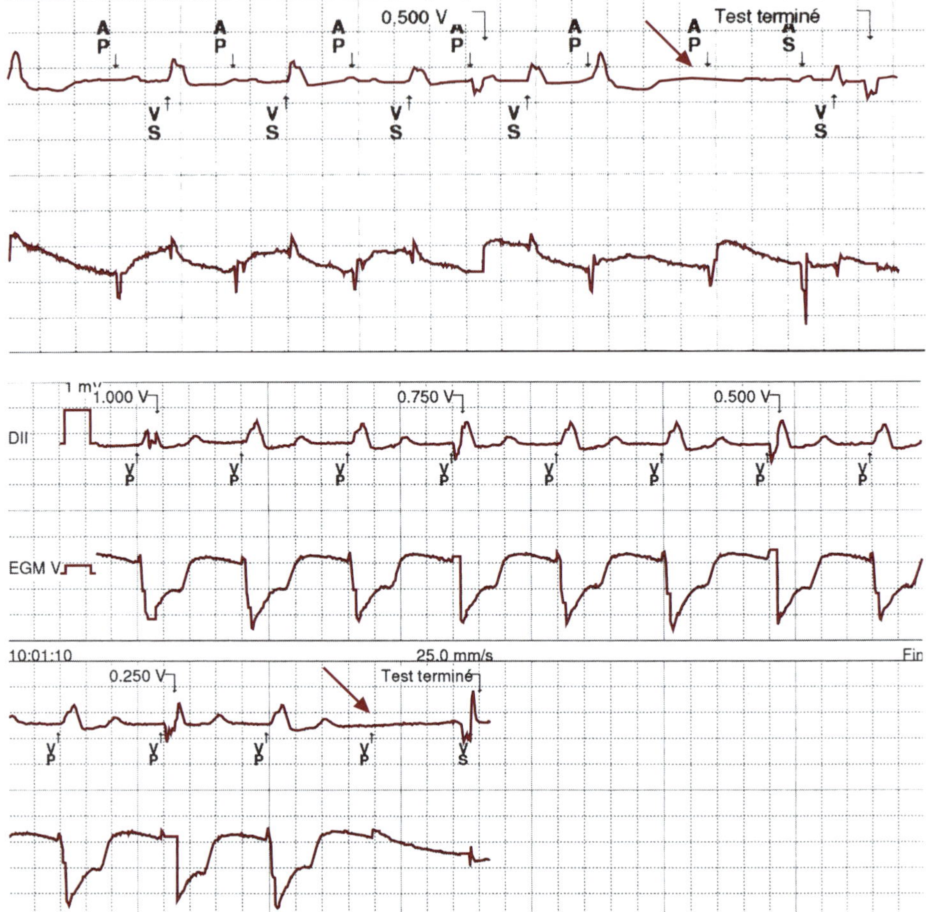

Figure 11.3 (a) Testing for atrial capture. A paced atrial stimulus with atrial capture will always be followed by a native ventricular stimulus (Vs) in the presence of normal AV conduction, even if no atrial stimulation artifacts are clearly visible. Decreasing the output to 0.5 V leads to a paced stimulus (arrow) without activation artifact and no ensuing Vs; the threshold is then established to be 0.75 V (for the given pulse witdh).
(b) Testing for ventricular capture. With progressive diminution of the ventricular output, loss of capture (arrow) occurs at 0.25 V; the stimulation threshold is then 0.5 V (for the given pulse witdh).

reviewed every 6 months or less, according to remaining battery energy and manufacturer's recommendations.

IV. BASIC PROGRAMMING

The basic function of a pacemaker is to detect native events and to pace whenever needed; this is accomplished by having a programmable pacing rate (lower and

Table 11.6. Indications of Pacing in the Coronary Sinus

- Absence of acute ST elevation
- Predominantly positive or biphasic ECG (with RBBB morphology) and inferior axis
- High pacing thresholds
- Atrial or simultaneous atrial and ventricular pacing
- Absence of ventricular ectopy with electrode manipulation
- Posterior course of the electrode on fluoroscopy or a lateral chest x-ray
- Recording both atrial and ventricular EGMs from the electrode

Table 11.7. Steps in Reviewing a Patient after Pacemaker Implantation

- Focused history taking (e.g., recurrence of preimplant symptoms)
- Focused physical examination (scar, generator pocket)
- Full medication review
- Magnet rate, battery voltage and impedance, pacing lead impedance
- Atrial/ventricular sensing and pacing thresholds
- Review of programmed parameters
- Verifying of pacemaker diagnostics (% of pacing, detected arrhythmias, mode switches, etc)
- Comparison of obtained values with previous checks, optimizing settings if necessary/possible
- Troubleshooting if needed

maximum) and an AV interval (for dual-chamber systems), which tries to mimic the PR interval. Optimization of this basic functioning is achieved by programmable sensing and pacing thresholds, various RPs and tracking rates, rate modulation, and so forth. A typical dual-chamber system programming is found in Figure 11.4.

A. Pacing Rate

A pacing rate determines the limits within which the pacemaker will pace the heart. Unless malfunctioning, a pacemaker should not allow the HR to go below its set lower pace rate; uncommonly, programming features may allow this to happen: hysteresis permits a lower rate between sensed events (while the paced rate is higher); algorithm terminating a pacemaker-mediated tachycardia; PVC response (automatic PVARP extension post-PVC to avoid triggering a pacemaker-mediated tachycardia [PMT]); atrial-based timing system (here a PVC resets the AA interval, and not the atrial escape interval as the ventricular-based systems do, and then adds the AV

Modes

Mode	AAIR<=>DDDR
Mode Switch	On
Detection Rate	175 bpm
Detection Duration	No Delay
Blanked Flutter Search	On

Rates

Lower Rate	60 ppm
Upper Tracking Rate	130 ppm
Upper Sensor Rate	130 ppm
ADL Rate	110 ppm

Intrinsic/AV

Paced AV	150 ms
Sensed AV	120 ms
Rate Adaptive AV	Off

Refractory/Blanking

PVARP	Auto
Minimum PVARP	250 ms
PVAB	180 ms
Ventricular Refractory	230 ms
Vent. Blanking (after A. Pace)	28 ms
PMT Intervention	Off
PVC Response	On
Ventricular Safety Pacing	On

Atrial Lead

Acute Phase	Off
Acute Phase Complete	01/27/09

Ventricular Lead

Amplitude	2.000 V
Pulse Width	0.40 ms
Sensitivity	5.60 mV
Sensing Assurance	On
Pace Polarity	Bipolar
Sense Polarity	Bipolar
Lead Monitor	Adaptive
Maximum Impedance	4,000 ohms
Minimum Impedance	200 ohms
Monitor Sensitivity	8
Capture Management	Adaptive
Amplitude Margin	2x
Min. Adapted Amplitude	2.000 V
Capture Test Frequency	Day at Rest
Acute Phase	Off
Acute Phase Complete	01/27/09
V. Sensing During Search	Adaptive

Additional/Interventions

RDR Detection Type	Off
Sleep	Off
Non-Comp. Atrial Pacing	On
Transtelephonic Monitor	Off

Figure 11.4. Typical basic programming of a dual-chamber pacing system. © Medtronic, Inc. 2010.

Rate Response

Optimization	On
ADL Response	4
Exertion Response	3
ADLR Percent	4.0%
Activity Threshold	Low
Activity Acceleration	30 sec
Activity Deceleration	Exercise
High Rate Percent	0.2%
ADL Rate Setpoint	8
Upper Sensor Rate Setpoint	19

Atrial Lead

Amplitude	1.500 V
Pulse Width	0.40 ms
Sensitivity	0.50 mV
Sensing Assurance	On
Pace Polarity	Bipolar
Sense Polarity	Bipolar
Lead Monitor	Adaptive
Maximum Impedance	4,000 ohms
Minimum Impedance	200 ohms
Monitor Sensitivity	8
Capture Management	Adaptive
Amplitude Margin	2x
Min. Adapted Amplitude	1.500 V
Capture Test Frequency	Day at ...
Capture Test Time	1:00 AM

Extended Telemetry	Off
Extended Marker	Atrial Sense Margin
Implant Detection	Off/Complete
Conducted AF Response	On
Maximum Rate	110 ppm
Post Mode Switch Pacing	Off
Atrial Preference Pacing	Off

Atrial High Rate Episodes

Episode Trigger	Mode Switch
Detection Rate	175 bpm
Detection Duration	No Delay
Collection Delay	30 sec
Episode Collection Method	Rolling

Ventricular High Rate Episodes

Detection Rate	180 ppm
Detection Beats	5 beats
Termination Beats	5 beats
SVT Filter	On
Episode Collection Method	Rolling

Figure 11.4. (*Continued*).

interval, so the interval from the sensed R wave to the next paced ventricular beat exceeds the lower rate interval, which is a form of compensatory pause).

In single-chamber systems the basic escape interval starts with a paced or sensed event and times out at an interval equivalent to the lower pacing rate; if no events are sensed at this point, a pacing stimulus will be triggered. In dual-chamber systems the escape interval has 2 components: the atrial escape interval starts with a sensed or paced ventricular event, followed by an atrial stimulus when it times out; the AV interval goes from the atrial event (sensed or paced) to the ventricular pacing stimulus.

Because tracking of P waves could result in very high ventricular rates in the presence of atrial tachyarrhythmias, there is a programmable limit, the maximum tracking rate, which is the maximum paced ventricular response or the shortest interval initiated by a sensed atrial event at which a paced ventricular event can follow a preceding paced or sensed atrial event. The maximum tracking rate is always defined by the total atrial refractory period (TARP) (see following for details) and its timer, initiated by a paced or sensed ventricular event, and must be completed before another ventricular pacing stimulus is triggered: this happens this way even when a P wave occurs and the subsequent AV interval times out.

B. AV Interval

The programmed AV interval has 2 components:

- **Blanking period (12–50 milliseconds):** Here all sensing is suspended, even the intrinsic ventricular waves, to avoid ventricular sensing of the leading edge of the paced atrial stimulus. If the blanking period is too long (> 100 milliseconds) an intrinsic R wave occurring soon after the atrial stimulus will not be sensed and a competitive ventricular stimulus will be delivered; if the blanking is too short there could be crosstalk and resulting ventricular inhibition.
- **Crosstalk sensing window:** To avoid inhibition of pacing if atrial activity is sensed by the ventricular channel, a ventricular-paced stimulus may be delivered 100 to 110 milliseconds after the atrial event (when the ventricular safety pacing feature is activated).

Variations in duration of the normal AV interval (125–200 milliseconds) may result from normal programming: (1) ventricular safety pacing interval (usually set at 110 milliseconds), (2) differential AV interval (the interval is 20–50 milliseconds shorter after a sensed atrial event than after a paced atrial event), (3) rate-adaptive AV interval (shorter intervals as atrial rate increases), and (4) AV interval hysteresis, which is periodic variations in interval either to maintain nodal conduction (positive hysteresis) or to avoid intrinsic conduction (negative hysteresis).

A short AV interval decreases the TARP and increases the atrial sensing window, which can result in atrial tracking at higher rates.

C. Refractory Periods

RPs in pacemaker programming have 2 different components: absolute RP is the blanking period, created to avoid oversensing; in the relative RP, events can be sensed but they are not used to trigger or reset pacing stimuli. This period is used to detect rapid signals: if they are at > 400 to 600 cycles/min (above the physiologic range), then electrical noise is likely and the pacemaker goes into asynchronous behavior (Noise Mode Response). If the rapid events are detected in the atrial channel the most likely reason will be an atrial tachyarrhythmia and the pacemaker may go into Automatic Mode Switch (see following).

An RP of particular importance is the TARP, formed by the AV interval and the postventricular atrial refractory period (PVARP). The PVARP is an RP that is initiated on the atrial channel by a sensed or paced ventricular event (to avoid far-field sensing), while the AV interval also works as an RP on the atrial channel of the pacemaker. The timing of native atrial events in relationship to the duration of the TARP determines which atrial events will be sensed and which will not, which may lead to abnormal pacing (upper rate behavior) in cases of rapid atrial tachyarrhythmias. Within the TARP is another RP (an independent timing interval), the postventricular atrial blanking (PVAB), which is an absolute RP triggered by ventricular paced or sensed events in order to prevent far-field sensing of ventricular signals on the atrial channel.

D. Rate Modulation

The addition of sensors (identified by the R word in the fourth position of the NPG code) permits modulation of the HR by the sensor input; this feature is of particular importance in patients with SND—for example, chronotropic incompetence, when the HR does not increase in response to physiological stress. Among several designs, sensors based on acceleration are most commonly utilized; their programming often requires some "fine touching" to better approximate the sensor's reaction to the needs of individual patients. Because accelerometers only respond to changes in frequency and intensity of motion, the concomitant use of respiration sensors (e.g., minute ventilation) permits better response to different workloads (like walking up stairs).

Because sensor-modulated HR changes do not inhibit native modulation by tracking of sensed P waves, several interactions may, and do, occur, mostly in the realm of upper rate behavior and undersensing/loss of sensing (most pacemakers and programming software now have special features incorporated that help avoid inappropriate combinations).

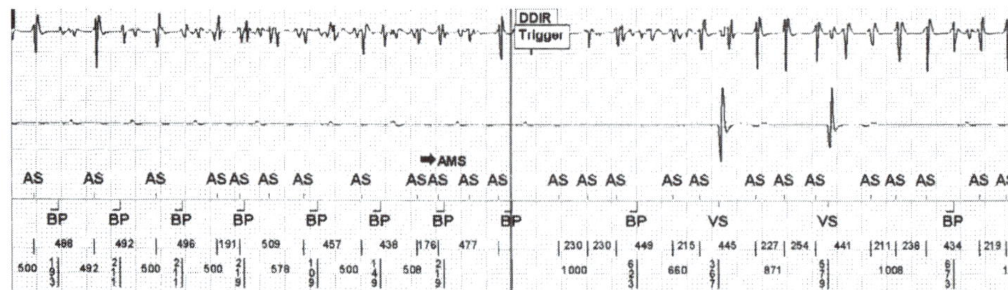

Figure 11.5. Automatic mode switch. Upon recognition of an atrial tachyarrhythmia with fast AV conduction the pacemaker switches automatically to a nonatrial tracking mode (DDIR), which maintains ventricular pacing if necessary.

E. Selected Associated Features

Rate drop response is a refinement of search hysteresis, in which intrinsic events detected below a programmed lower rate trigger more rapid-paced stimuli. This feature may be of use in patients with cardioinhibitory carotid sinus syndrome.

Automatic mode switch is another feature. In the presence of atrial tachyarrhythmias the pacemaker reprograms itself automatically to a mode (e.g., VDIR, DDIR) that no longer follows the intrinsic atrial rate (Figure 11.5). This is possible because the pacemaker detects atrial events occurring within the PVARP but does not track them.

The automatic mode switch feature can be enhanced by increasing atrial sensitivity (optimal is 3 times the safety margin, compared with 2 times for sinus P-wave sensing), decreasing the mode switch rate, or decreasing the PVAB.

F. Pacemaker Diagnostics

Apart from allowing verification and modification of programmed parameters, modern pacemakers are increasingly interactive, with self-diagnostic capabilities and enhanced telemetry. Automatic sensing and capture threshold measurements are possible in both atrial and ventricular leads. Event markers, also in both atrial and ventricular channels, are indicators of sensing and pacing as reported directly by the pacemaker; they are most useful when simultaneously correlated with the surface ECG. Intracardiac EGMs help determine the morphology of native complexes and troubleshoot sensing/capture problems.

V. PACEMAKER COMPLICATIONS AND TROUBLESHOOTING

Pacemakers need regular monitoring to ensure adequate functioning and to detect possible malfunctions. Their complications can be broadly divided into those related to system malfunction, including functional complications, and those that are not.

A. Complications Not Related to System Malfunction

1. Infections

A pacemaker could become infected at the level of the generator pocket or deeper (intravenous lead, endocardial). This incidence ranges from 0.8% to 5.7%. Commonly cited risk factors are the presence of diabetes, operator inexperience, recent treatment with glucocorticoids or anticoagulants, underlying malignancy, and prior surgery involving the device (the risk for infection is 4 times higher at the time of generator replacement).

The germs most commonly detected are *S. Aureus* and coagulase-negative staphylococci (such as *S. Epidermidis*), causing 65% to 75% of generator pocket infections and up to 89% of device-related endocarditis. The use of prophylactic antibiotics prior to pacemaker implant has a protective effect. The presenting symptoms are usually fever and chills, with local signs of the infection or signs of device-related endovasculitis/endocarditis. Erosion through the skin of the generator or a portion of a lead is seen occasionally but it is not necessarily an indication of infection (though it should always be suspected) as it can also be caused by local pressure or trauma.

Treatment of pacing system infection includes antibiotic coverage, explantation of the generator and leads (with temporary pacing in patients pacing-dependent), and subsequent reimplantation of a new system once the infection has been totally controlled. The reimplant is usually carried out in the site contralateral to the side of the infection.

2. Bleeding

Active bleeding or hematomas of the generator pocket are occasionally seen, mainly in those who are on chronic oral anticoagulation and received IV heparin. Depending on the severity of the bleed, a compressive bandage of the site with periodic surveillance may suffice, but surgical evacuation with appropriate hemostasis is occasionally required. The clinician should always resist the impulse to drain a hematoma percutaneously as this rarely resolves the cause of the bleeding and does increase the risk of infection. The need for continuing anticoagulation should always be reassessed.

3. Thrombosis

Thrombosis related to transvenous leads occurs with relative frequency (23% 1-year cumulative incidence in one study); it is rarely related to significant local morbidity and can be a source for pulmonary embolism. Known prothrombotic risk factors are use of oral contraceptives, hormone replacement therapy, and personal history of venous thrombosis. Thrombosis is most common in the first 3 months postimplant and in many instances does not cause clinical symptoms. When symptomatic, pain, discoloration, local swelling or edema, and visible collateral circulation are frequently seen. The diagnosis is usually confirmed with Doppler ultrasonography or, less frequently, a contrast venography. Treatment of local thrombosis includes IV heparin (or LMWH) followed up by oral anticoagulation; very symptomatic patients may benefit from thrombolytic therapy, suction thrombectomy, or, very rarely, surgery.

4. Pacemaker syndrome

Pacemaker syndrome is a constellation of symptoms resulting from loss of AV synchrony due to single-chamber VVI pacing. The resulting decrease in stroke volume leads to presyncope/syncope, easy fatiguability, cough, dizziness, dyspnea on exertion, orthopnea, paroxysmal nocturnal dyspnea, and so forth. Physical examination may reveal arterial hypotension (typically VVI pacing drops BP > 20 mm Hg), signs of HF, and murmurs of tricuspid and/or mitral regurgitation. Management of pacemaker syndrome includes: (1) upgrading to a DDD system, (2) decreasing lower pacing rate to encourage generation/conduction of sinus beats, (3) use of hysteresis, and (4) withdrawal of medications that impair sinus node function. In the rather rare cases of pacemaker syndrome in dual-chamber systems, programming should be optimized to ensure atrial capture; atrial nonpacing modes (e.g., VDD) and atrial nontracking modes (e.g., DDI, DVI) should be avoided. Fortunately, pacemaker syndrome is a rather unusual problem and most patients tolerate well asynchronous pacing.

5. Tricuspid regurgitation

Occasionally, pacing leads across the tricuspid valve may rend it dysfunctional (due to perforation, entanglement, impingement, or adherence to the tricuspid leaflets), most commonly resulting in varying degrees of regurgitation. Severe tricuspid regurgitation may lead to severe HF. Though most patients will not suffer an aggravation of the degree of regurgitation, symptomatic cases should be managed appropriately, uncommonly going as far as lead explantation with implantation of pericardial leads instead, and repair/replacement of the tricuspid valve.

B. Complications Related to System Malfunction

1. Output and capture

a. Failure to output—no pacing stimulus (atrial or ventricular) is documented in the ECG: This is a normal finding if the patient has a native rhythm and no pauses longer than the programmed lower rate of the pacemaker are present (unless the hysteresis function is "on"). Otherwise, several reasons may explain the absence of pacing stimuli (Table 11.8).

Once the replacements indicators are "on" (RRT or ERI), there is usually 3 to 6 months of adequate functioning before complete loss of output occurs. Rarely, circuit failure may lead to failure to output: replacing the generator is the only treatment option.

b. Failure to capture—the ECG displays no depolarizations after atrial or ventricular pacing stimuli at a time when the myocardium is not in the RP: Loss of capture may occur abruptly or due to progressively increasing pacing thresholds (Figure 11.6); it may also be total or intermittent.

Table 11.8. Failure to Output and Troubleshooting Options

Causes	Characteristics	Troubleshooting Options
Oversensing of inappropriate electrical signals	Unexpected sensing markers are present on the marker channel of the pacemaker Pacing resumes when magnet is placed over the pacemaker	Decrease sensitivity Change mode to DVI or VVI Perform invasive test and treat accordingly
Open circuit (loose set screw, lead not fully inserted, insulation break, fractured lead wire, air in pocket)	Pacing markers are present on the marker channel Magnet use fails to eliminate the problem Confirm with chest x-ray, fluoroscopy, lead tests	Reprogram pacing to unipolar Perform invasive test and treat accordingly
Battery depletion	Previous documentation of recommended replacement time (RRT) or elective replacement indicator (ERI), with changes in the magnet-induced rate and pacing mode No pacing markers are documented Interrogation of the pacemaker is often not possible	Replace the unit

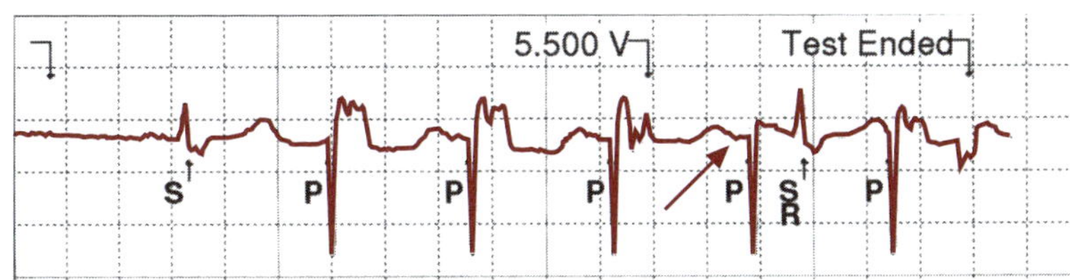

Figure 11.6. Increased pacing threshold at 6 V; a pacing stimulus (arrow), with no activation artifact, along with sensing of a native beat (SR), is noted when voltage is decreased to 5.5 V.

There is a close correlation between many causes of noncapture and the timing of the implant (Table 11.9).

In all situations, electrolyte imbalances (e.g., hyperkalemia) and drug effects (e.g., AAD) should be excluded and, if present, managed accordingly before invasive lead testing is carried out.

Table 11.9. Failure to Capture and Troubleshooting Options

Timing of the Implant	Causes	Characteristics	Troubleshooting Options
Within hours or days after implant	Lead dislodgement or malposition	Change in morphology of the pacing-evoked depolarizations when capture is present Change in lead position in chest x-ray Associated undersensing as well as atrial or ventricular ectopy	Reposition the lead
	Inadequate safety margin	Capture occurs with increased V and/or PW Stable position of the lead in a chest x-ray	Program PW and/or V to higher value
Within weeks to months	High capture thresholds due to lead maturation	Capture occurs with increased V and/or PW Stable morphology of the pacing-evoked depolarizations when capture is present Stable position of the lead in a chest x-ray	Program PW and/or V to higher value and reassess (threshold will improve over time) Reposition lead as needed
Months to years	Exit block	Capture occurs with increased V and/or PW Stable morphology of the pacing-evoked depolarizations when capture is present Stable position of the lead in a chest x-ray	Program PW and/or V to higher value Reposition lead as needed

Table 11.9. *(Continued)*

Timing of the Implant	Causes	Characteristics	Troubleshooting Options
	Structural problems with the lead	*Insulation break:* low impedance ($< 250\ \Omega$), high current drain, undersensing, extracardiac stimulation possible *Lead wire fracture:* high impedance ($> 1200\ \Omega$, low current drain (may be normal or erratic if partial fracture), no sensing (may be low and erratic if partial fracture), point of fracture in chest x-ray	Reprogram pacing to unipolar while awaiting replacement of the lead
	Structural problems in the myocardium (focal necrosis, progressive fibrosis)	Capture may occur with increased V and/or P W Associated clinical signs and symptoms	Reprogram pacing to unipolar while awaiting repositioning of the lead

Change up to 300 Ω in lead impedance may occur as a result of lead maturation; it is considered within normal limits and may still be compatible with normal lead function.

If bipolar pacing fails to capture but unipolar pacing is good, the reason could be a loose anodal setscrew.

If a lead needs to be explanted, simple traction is usually good enough for newly implanted electrodes (though there is always a risk of endocardial/vascular trauma and lead damage), but in chronic implants the use of specialized extraction techniques (stylets, radiofrequency energy, or laser) by adequately trained personnel is necessary. Depending on the clinical situation many physicians would prefer to leave the old lead in place and just implant a new one.

2. Sensing

a. Oversensing—inappropriate pacemaker inhibition below the programmed base rate, or presence of sensing markers on the marker channel without native depolarizations on the ECG: Oversensing results from detection of inappropriate signals, physiological or nonphysiological, or from structural problems of the leads (Table 11.10). It occurs more commonly in unipolar configurations because the dipoles are located far apart (generator and tip of the lead) while bipolar systems have the dipoles located nearby in the lead, which results in less (but still possible) interference.

Table 11.10. Oversensing and Troubleshooting Options

Causes	Characteristics	Troubleshooting Options
Sensing of inappropriate electrical signals (myopotentials, EMI, far-field sensing, connector problem, partial wire fracture) Sensing circuit failure	Normal lead appearance on chest x-ray Sense markers on the marker channel of the pacemaker, without native depolarizations on ECG Sensing may resume in unipolar leads with device manipulation or patient's exercise	Decrease sensitivity Increase RP if far-field sensing Change mode to DVI or VVI If needed, perform invasive test and treat accordingly No change may be option in unipolar leads if no other problems are found
Lead wire fracture Lead partially inserted	Abnormal lead appearance on the chest x-ray Abnormal lead tests	Change mode to DVI or VVI Perform invasive test and treat accordingly

Sensing of far-field R or T waves by the atrial lead, which may trigger inappropriate automatic mode switch, can be solved by decreasing its sensitivity or by increasing the PVARP.

A peculiar situation with ventricular oversensing is the presence of a sequence of atrial pacing followed by both a ventricular-sensed and a ventricular-paced stimulus, all within 110 milliseconds. This pattern is highly suggestive of ventricular safety pacing, but it may also be due to electromagnetic interference, sensing of myopotentials, atrial afterpotentials, or a too high sensitivity. Programming the ventricular safety pacing to "off," removing the source of electromagnetic interference, testing for myopotentials, and/or reprogramming the sensitivity (to a higher value, decreasing sensitivity) should resolve the problem.

In dual-chamber systems ventricular oversensing results in rate decreases below the programmed base rate; if a ventricular RP and PVARP are triggered by the sensed event, functional atrial and ventricular undersensing may occur. Atrial oversensing leads to increases in rate (myopotential drive) with brief periods of ventricular pacing at or near the maximum tracking rate.

When the hysteresis function is activated, there could be pauses longer than the programmed base rate: it is recognized when the pacemaker remains inhibited in between native heart beats lower than the programmed base rate, or when pauses occur only after native sensed beats.

b. Undersensing—presence of pacing stimuli inappropriately associated with native depolarizations, or native depolarizations on the ECG without corresponding sensing markers on the marker channel: Diagnosing undersensing in the absence of sensing markers on the marker channel of the pacemaker, while native depolarizations are documented on the ECG, is a fairly straightforward exercise (Figure 11.7). Troubleshooting options are shown in Table 11.11.

Difficulties in determining true undersensing are encountered when no EGM markers are available, and the ECG shows pacing stimuli coinciding with ectopic

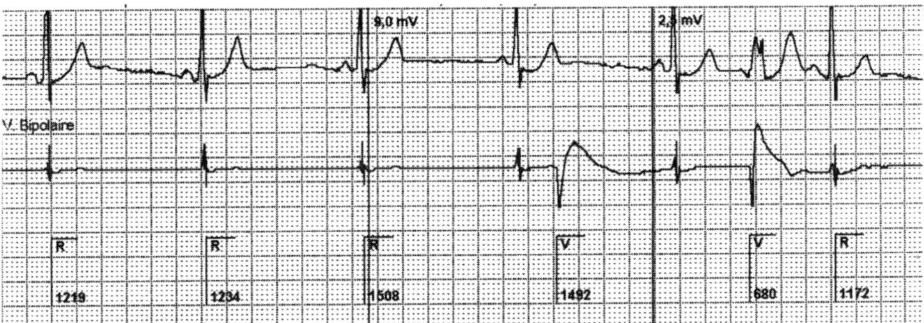

Figure 11.7. ECG tracing showing undersensing. The fourth and fifth native beats are not sensed, and artificial stimulation (V) ensues.

Table 11.11. Undersensing and Troubleshooting Options

Causes	Characteristics	Troubleshooting Options
Lead wire fracture Lead dislodgement Insulation break	Abnormal lead appearance on chest x-ray (may be normal in some cases) Abnormal lead test	Change mode to DVI, VVI, or VOO, increasing rate as needed, while awaiting replacement/repositioning of the lead
Lead maturation Inadequate signal (low amplitude, slew rate, or frequency content) Inadequate sensing margin Electrolyte imbalance Drug effect	Normal lead appearance on chest x-ray Normal lead test Sensing may improve with increased sensitivity	Increase sensitivity Restore electrolyte balance Modify drugs Reposition lead as needed
Sensing circuit failure	Normal lead appearance on chest x-ray EGMs are appropriate signals for sensing Normal lead test Reprogramming does not modify the results	Increase rate and program to VOO while awaiting replacement of the generator

beats (the extra stimulus may not be sensed and a pacing stimulus is delivered), in which case no real abnormality is present, or when pacing stimuli coincide with regular native pacemaker depolarizations, in the presence or absence of AV block (Table 11.12).

The physician should always be aware that magnet application and noise interference suspend the sensing function and force the pacemaker into asynchronous function.

Among complications related to system malfunction, device advisories/recalls occupy a particular place. Despite significant progresses in pacing technology, failures are bound to happen: they range from hardware problems (generator failure, premature battery depletion) to software glitches (telemetry problems, inappropriate signal detection, etc). Even though there are systems in place that alert the physician and public, centers involved in pacemaker implantation/follow-up should have established policies in dealing with these issues. Specific recommendations on

Table 11.12. ECG Characteristics during Undersensing

AVN Conduction	ECG Characteristics	
	Loss of Atrial Sensing	Loss of Ventricular Sensing
Complete AV block	Sequential AV pacing at the programmed base rate of the pacemaker Native P waves "marching" through the tracing, with occasional resetting (due to atrial capture) and functional noncaptures (due to pacing stimuli in the atrial RP)	Not obvious because each P wave is tracked and triggers a ventricular pacing stimulus
Normal AVN conduction	Difficult recognition Several rhythms possible	Recognition is difficult and possible only if the programmed AV interval is long: the pacing stimulus occurs well after the native QRS

how to deal with advisories/recalls usually come from manufacturers, and they may include software/programming changes, increased follow-up, and in some cases device extraction/replacement.

3. Functional system malfunction

Functional malfunctions are not true pacemaker malfunctions: they mostly occur when sensing and pacing events coincide with the RP or when pacemaker-induced rhythm disorders are present.

a. Functional oversensing: Functional oversensing may be due to short blanking periods, RP extensions, and latency between the surface ECG and intrinsic conduction. Functional atrial oversensing may be prevented by (1) programming the automatic PVARP extension after PVC to "off," (2) eliminating the PVARP extension whenever a P wave is detected within the PVARP preceding a sensed ventricular event, and (3) activating the noncompetitive atrial pacing feature.

b. Functional noncapture: A functional noncapture develops when the pacing stimulus coincides with the RP, due to a failure to sense or to functional undersensing; in all cases its treatment is aimed at the underlying problem.

c. Pacemaker-induced rhythm disorders:

(1) Crosstalk: Due to far-field sensing in the opposite channel, most commonly ventricular sensing of atrial stimuli. Crosstalk is seen in DDD, DVI, and DDI pacing modes, and it is suggested by failure (inhibition) of ventricular pacing following paced atrial events but not after atrial sensed events. If the ventricular safety pacing is "on," then the AV interval will be shorter than programmed (110 milliseconds). Crosstalk can be managed by decreasing atrial output, decreasing ventricular sensitivity, programming pacing mode to VDD, increasing the crosstalk sensing window, and asynchronous pacing.

(2) High-rate atrial tracking: Due to pacemaker tracking of an SVT or atrial oversensing, both resulting in a paced tachycardia at or near the programmed maximum tracking rate. Activation of the automatic mode switch feature should terminate it in cases of SVT; atrial oversensing should be managed as previously explained.

(3) Inappropriate automatic mode switch: Most commonly due to sensing of ventricular signals by the atrial lead; if it occurs during the PVARP it means that the PVAB is too short or the QRS is too long. Its management includes increasing the PVAB period, discontinuing medications that may prolong the QRS (flecainide, amiodarone), treating hyperkalemia (because it also prolongs the QRS), and decreasing atrial sensitivity if the PVAB period is not programmable.

Inappropriate mode switch may also be due to double counting of P waves (near-field oversensing), or sensing of the early part of native QRS (far-field oversensing) within the AV interval. In all cases, if the mode switch needs to be reprogrammed, a rate that is less than the arrhythmia rate but higher than the maximum tracking rate is to be chosen.

(4) Upper rate behavior: Normally the interval between consecutive native atrial events is longer than the TARP, with each P wave occurring in the atrial alert period being therefore sensed; if the native PP interval happens to be shorter than the TARP then some P waves will not be sensed because they will fall in the TARP. As a consequence, Wenckebach-like behavior or abrupt fixed block may occur.

Wenckebach-like behavior (electronic Wenckebach): In dual-chamber systems, as previously explained, the maximum tracking rate timer must be completed before a ventricular pacing stimulus is released, even when a P wave occurs and the subsequent AV interval times out. This delay lengthens the interval from the sensed P wave to the ventricular pacing stimulus (PV interval) while placing the next sensed P wave closer to the preceding ventricular-paced complex. If this cycle persists, a P wave will eventually coincide with the PVARP and will not be sensed or tracked, resulting in a relative pause. The clinical effect is fluctuation in rate (group beating) at a fixed ventricular rate, progressively lengthening PV interval and intermittent pauses. The patient may become symptomatic, mostly with palpitations. This problem is usually solved by increasing the maximum tracking rate.

Abrupt fixed block occurs when only every other P wave is sensed (the other falling in the TARP); because the pacemaker tracks only the sensed events, a fixed 2:1 block develops (higher levels of block may occur depending on the native atrial rate). By dividing 60,000 milliseconds by the TARP one determines the frequency at which the block may occur (so never programming a maximum tracking rate above this number is the easiest way to avoid this type of dysfunction). Associated symptoms (presyncope, sudden fatigue) are secondary to the drastic drop in rate. This anomaly is treated by modifying the TARP (either by decreasing the PVARP or the AV interval), or by programming countermeasures (rate-smoothing, fallback). Abrupt 2:1 block may sometimes occur in those patients who can exercise to a sufficient degree to achieve high atrial rates.

(5) Pacemaker-mediated tachycardia: Endless loop tachycardia is the commonest form of PMT, due to atrial sensing of retrograde P waves: the pacemaker acting as the antegrade limb of the arrhythmia circuit while the normal conducting system acts as the retrograde component. PMT requires AV dissociation to start it (its cause should always be searched) or a pacing stimulus that fails to depolarize the atrium and leaves it amenable to retrograde depolarization. The ECG shows intermittent or continuous ventricular pacing at or near the upper rate limit. The best way to treat PMT is to program a long PVARP so that the pacemaker does not sense or track retrograde P waves; when doing this the physician should be aware that increasing the PVARP increases the TARP, which subsequently limits the maximum atrial rate that can be sensed and tracked. If a reduced maximum atrial rate is not advisable, other possible therapeutic options are the use of a magnet to suspend most programming and revert the pacemaker to basic functioning (temporary measure), use of CSM or AVN-blocking drugs, programming of automatic PVARP extension following a PVC, "a pace on PVC" feature in some models, and reprogramming to nonatrial tracking modes (DDI, DVI, VVI).

VI. PHYSIOLOGIC PACING

Placement of a lead at the apex of the RV is relatively simple and provides the required lead stability, but chronic and frequent RV apical pacing can be detrimental for LV function. The resulting dyssynchronous LV contraction leads to impairment in systolo-diastolic function with secondary myocardial perfusion abnormalities, remodeling, pacemaker syndrome, and AF; clinical trials have also shown an overall increase in morbidity and mortality. Attempts at a more physiologic pacing are 2-pronged: minimization of unnecessary RV apical pacing and alternative pacing sites.

A. Minimization of Unnecessary RV Apical Pacing

This works through modification of the pacing mode:

- AAIR, DDI, DDIR pacing modes in patients with normal AV conduction.
- VVI with low programmed lower rate, usually 40 bpm, in patients needing ICD for primary prevention and no other pacing indications.
- Modifying the programmed AV delay to longer paced and sensed AV intervals, eliminating the rate-responsive delay, or using the AV hysteresis function (which allows automatic lengthening of the programmed AV interval if a native AV conduction is detected).
- Using dedicated pacing algorithms intended to minimize ventricular pacing (e.g., AAIsafeR, Managed Ventricular Pacing), which basically switch the pacemaker from an atrial-based pacing mode to a dual-chamber mode when the device detects an AV block.

B. Alternative Pacing Sites

Pacing the RV outflow tract, mostly its septal portion, is showing promising results in clinical studies. Direct His pacing, potentially the closest to intrinsic AV conduction, remains very attractive but needs a dedicated system to achieve technical success.

Multisite pacing (biventricular, dual-site univentricular) and single-site LV pacing have shown some benefit in initial studies but are technically more challenging, with increased risk of complications.

VII. BIVENTRICULAR PACING IN SYSTOLIC HEART FAILURE

A. General Principles

The utility of atrial-synchronous biventricular pacing in HF patients is based on the premise that stimulating late-activated regions of the LV restores contractile coordination and provides increased pumping effectiveness without increasing myocardial oxygen consumption; it has been showed to improve LV function, decrease symptoms, prolong survival, and increase quality of life in patients with advanced systolic HF. The effect of cardiac resynchronization therapy (CRT) comes mostly by delaying segments with early systolic contraction during sinus rhythm, but its effectiveness has also been proven in patients with permanent AF and AV block.

When analyzed together, the benefits of resynchronization by pacing are numerous: pacing allows optimization of medical therapy, for example by permitting increases in β-blocker dosage; dual-chamber pacing permits optimization of the AV interval, which results in better AV synchrony and minimization of mitral regurgitation;

atrial pacing may reduce the frequency of AF; by pacing the LV, RV-induced ventricular dyssynchrony is avoided or limited, which results in improved cardiac function, HF symptoms, and survival.

B. Patient Selection

CRT is the only type of pacing indicated to treat systolic HF (Table 11.13). Many patients who fulfill ICD implantation criteria do so for CRT, which explains why most CRT systems are ICD-based.

Before considering a patient for CRT, optimal recommended medical therapy (β-blockers, ACEIs, aldosterone antagonists) should have been tried for a significant period of time, often for at least 3 months postrevascularization of CAD and for 6 to 9 months postdiagnosis of NIDCM.

Table 11.13. Recommendations for CRT in Patients with Severe Systolic HF

Class I

- For patients with LVEF ≤ 35%, a QRS duration ≥ 0.12 s, and sinus rhythm, CRT with or without an ICD is indicated for the treatment of NYHA functional class III or ambulatory class IV HF symptoms with optimal recommended medical therapy

Class IIa

- For patients with LVEF ≤ 35%, a QRS duration ≥ 0.12 s, and AF, CRT with or without an ICD is indicated for the treatment of NYHA functional class III or ambulatory class IV HF symptoms with optimal recommended medical therapy
- For patients with LVEF ≤ 35%, with NYHA functional class III or ambulatory class IV symptoms who are receiving optimal recommended medical therapy and who have frequent dependence on ventricular pacing, CRT is reasonable

Class IIb

- For patients with LVEF ≤ 35%, with NYHA functional class I or II symptoms who are receiving optimal recommended medical therapy undergoing implantation of a permanent pacemaker and/or ICD with anticipated frequent ventricular pacing, CRT may be considered

Class III

- CRT is not indicated for asymptomatic patients with reduced LVEF in the absence of other indications for pacing
- CRT is not indicated for patients whose functional status and life expectancy are limited predominantly by chronic noncardiac conditions

The use of echocardiography in patient selection and prediction of response to CRT is based on the facts that mechanical dyssynchrony is not necessarily related to electrical dyssynchrony, and that LV dyssynchrony, not necessarily indicated by a wide QRS complex or presence of LBBB, is a major predictor of response to CRT. Mechanical dyssynchrony does reflect electromechanical delay, either at the AV level (the delay between atrial and ventricular contractions leads to a decreased ventricular filling time, mitral regurgitation with late diastolic regurgitation, and atrial systole coinciding with early passive filling, which decreases LV filling), interventricular level (as with LBBB), and/or intraventricular levels.

Several echocardiographic techniques of utility in assessing mechanical dysynchrony have been described (Table 11.14).

Echocardiography also allows assessment of acute and long-term effects of CRT (such as improvements in LV EF, reduction in LV end-systolic and end-diastolic volumes, reverse remodeling, reduction in mitral regurgitation) as well as optimization of pacing parameters (mostly AV and VV intervals).

C. Basics of CRT Implantation Technique

Together with a standard 2-lead implant (RA and apex of RV), a third lead is positioned to pace the LV through the coronary sinus or via an epicardial approach. The target cardiac veins in the coronary sinus are the posterolateral, lateral, and antero-

Table 11.14. Selected Echocardiographic Predictors of CRT Response

Techniques		Utility in Determining:
M mode	Septal-to-posterior wall motion delay ≥ 130 ms	Intraventricular dyssynchrony
	Interventricular mechanical delay ≥ 40 ms	Interventricular dyssynchrony
Pulse-wave or color-coded TDI	Basal septal-to-lateral delay ≥ 65 ms	Intraventricular dyssynchrony
Tissue synchronization imaging	Color mapping of timing of peak systolic velocities	Intraventricular dyssynchrony
Strain imaging	Septal-to-posterior wall peak strain ≥ 130 ms	Intraventricular dyssynchrony
Real time tridimensional echocardiography	Comparison of regional volume-time curves	Intraventricular dyssynchrony

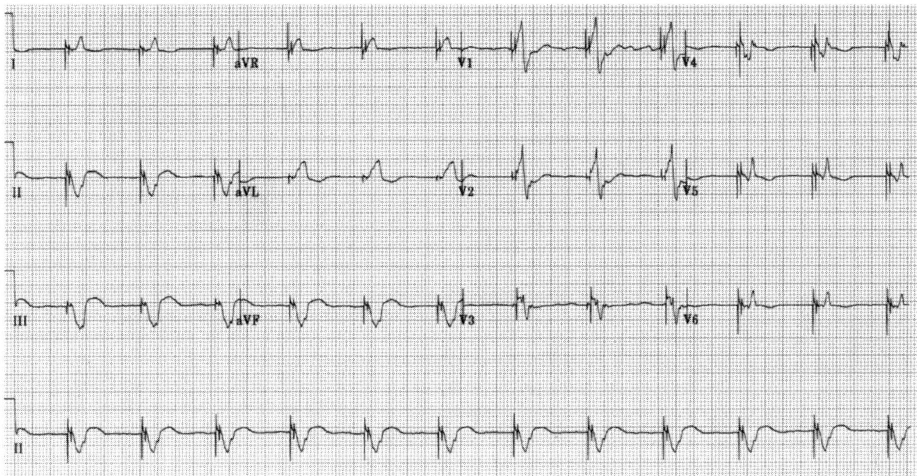

Figure 11.8. ECG showing biventricular pacing. The significant interventricular delay (40 milliseconds) allows the clear identification of the single ventricular stimulation spikes.

lateral veins (pacing from the anterior cardiac or the posterior interventricular veins stimulates primarily the septum and does not resynchronize the LV).

An acceptably good LV pacing threshold is < 2.5 V at 0.5 milliseconds of PW; occasionally the PW should be increased (up to 1.5 milliseconds) to improve capture. Because of the LV lead position, the implanting physician should always check for diaphragmatic stimulation during deep inspiration and high pacing output.

Because of the obvious biventricular nature of the pacing, the ECG morphology of CRT patients differs notably from RV-based pacing systems, mainly in the presence of narrower paced complexes and/or RBBB morphology (Figure 11.8). Apart from LV pacing, a paced QRS with RBBB morphology may be seen in cases of pseudofusion.

In cases of a failed transvenous implantation (usually due to failure to position the LV lead) a surgical approach is still feasible, by means of subxiphoid access, open thoracotomy, and video-assisted thoracoscopy, and even by using robotic technology. A transseptal approach has also been attempted.

D. Response to CRT

A good response to CRT is indicated by improvements in clinical parameters such as functional class or quality of life. Electrical and echocardiographic parameters are used to assess the response. Even though the QRS duration does not reflect LV dyssynchrony, a decreased QRS width (> 30 milliseconds) with pacing is considered a good marker of benefit. Among echocardiographic parameters, amelioration of intraventricular dyssynchrony is the best predictor of good response, but no specific

Table 11.15. Measures to Increase Response to CRT

Parameter	Approaches
LV lead placement	Lateral/posterolateral position, better than anterior/anterolateral Optimal lead position may also be determined with the help of tissue doppler imaging, searching for the latest contracting segment
Optimization of AV interval	Fixed short AV delay (~ 100 ms) Use of negative AV search hysteresis Complete AVN ablation in cases of AF Optimizing programming, e.g., activation of ventricular rate response
Ensuring biventricular pacing	Adequate pacing threshold, aided by "electronic repositioning" if necessary Optimal AV delay
Optimizing VV timing	LV offset between 0 and –20 ms seems best option for most patients

technique has been found to be superior at improving patient selection criteria or response rates.

When the response is assessed only on clinical grounds the prevalence of non-responders is around 30%; this figure increases to 40% when echocardiographic parameters are used. The key to a high probability of success with CRT relies on patient selection (see preceding), but there are ways to improve results in those not doing well (Table 11.15).

The ideal AV interval provides the longest LV filling time without premature truncation of the atrial wave by the mitral valve closure; by echocardiography, the end of the atrial wave should coincide with the onset of ventricular contraction. Because this assertion is not necessarily valid for all patients, apart from requiring time-consuming validation techniques, most physicians would just program a fixed short AV delay.

"Electronic repositioning" refers to noninvasive (via telemetry) changes in the LV stimulation configuration to improve the stimulation threshold or to avoid phrenic nerve stimulation. When using LV unipolar leads the possible configurations are only unipolar or extended bipolar (LV tip–RV coil, the latter acting as the anode), Bipolar leads allow various configurations: unipolar (stimulation from the proximal electrode only), extended bipolar (LV ring–RV coil), and dedicated bipolar pacing, with either electrode serving as the cathode (LV tip–LV ring, or LV ring–LV tip).

SUGGESTED READINGS

Ellenbogen KA, Wood MA. *Cardiac Pacing and ICDs.* 3rd ed. Boston, MA: Blackwell Science; 2002.

Epstein AE, Dimarco JP, Ellenbogen KA, et al. ACC/AHA/HRS 2008 guidelines for device-based therapy of cardiac rhythm abnormalities. *J Am Coll Cardiol.* 2008;51:2085-2105.

Chapter 12

Implantable Cardioverter Defibrillators

Miguel A. Barrero Garcia and Bernard Thibault

I. GENERAL PRINCIPLES

Since its initial conception in the 1960s and the first human implant in 1980, the implantable cardioverter defibrillator (ICD) has come a long way, evolving from a very basic "shock box" to a sophisticated device, with multiprogrammability, ease of implantation, and improved diagnostic and therapeutic capabilities. Able to stop ventricular tachyarrhythmias by means of overdrive pacing or by electrical shocks, multiple clinical trials assessing their use in both primary and secondary prevention of life-threatening cardiac arrhythmias have proven the utility of ICDs, making them an advisable therapy in selected patients (Figure 12.1).

A. Ventricular Fibrillation and Defibrillation

VF is the most commonly recorded rhythm at the time of sudden cardiac arrest, and its EP mechanism remains the focus of much active investigation. There are multiple proposed hypotheses based in animal models, including (but not limited to) the multiple-wavelet hypothesis, spiral-wave reentry, scroll waves, or focal repetitive activation. Regardless of the specific mechanism, VF is the final common pathway leading to SCD.

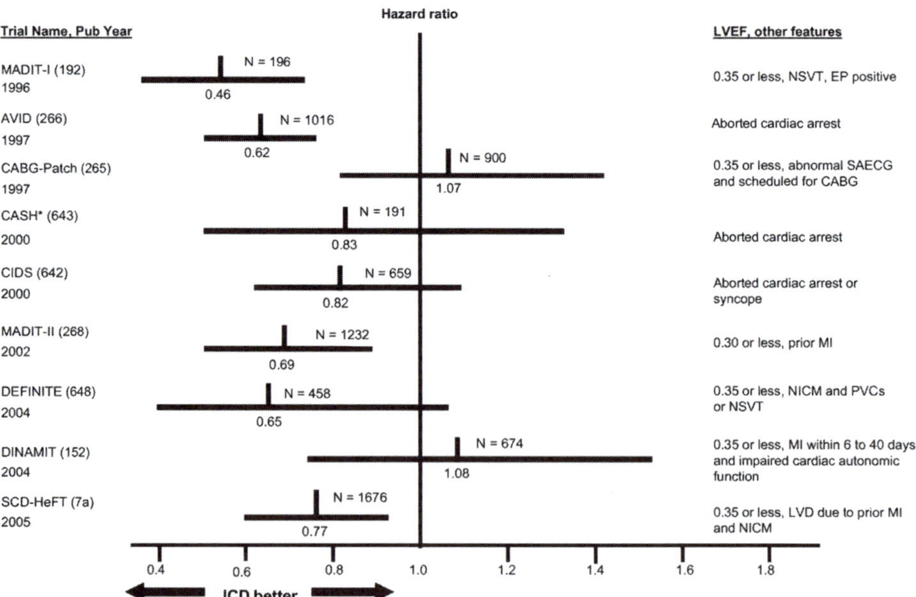

Figure 12.1. Major clinical trials on ICD therapy. Reprinted from Zipes DP, Camm AJ, Borggrefe M, et al. ACC/AHA/ESC 2006 guidelines for management of patients with ventricular arrhythmias and the prevention of sudden cardiac death. *J Am Coll Cardiol.* 2006;48:247-346, with permission from Elsevier.

Attempts to analyze VF have provided very useful information on how to terminate it by electrical defibrillation, which is achieved when a critical mass of myocardium is depolarized, interrupting the ongoing arrhythmia and preventing its reinitiation. The probability of a successful defibrillation depends on the energy of the shock, with each shock at any given energy level having a probability of success. The defibrillation threshold is defined as the shock strength that terminates VF in an individual 50% of the time; because defibrillation is a probabilistic phenomenon and taking into account the shape of t he defibrillation success curve (higher energies associated to higher success rate), it is assumed that energy delivery > 10 J above the defibrillation threshold (DFT) should defibrillate 99% of the time.

The defibrillation waveform, as the electrical impulse comes out of the ICD, is a truncated (terminated before full capacitor discharge) exponentially declining pulse, with energy delivery varying inversely with resistance (with fixed pulse). Waveforms in current ICDs are of biphasic nature because of their superior performance over the monophasic ones. This superiority seems to be better explained by the burping theory: the first phase of the biphasic shock depolarizes or extends the RP of most ventricular myocytes while the second phase removes excess charge from any cell where it remains.

The upper limit of vulnerability (ULV) is the weakest measured shock strength that does not induce VF when a shock is given in the vulnerable period of the cardiac cycle (the ascending part of the T wave); as such, it is a good correlate of the DFT (which is the lowest shock strength that terminates VF) because, if we consider that a successful defibrillation occurs when the shock fails to reinitiate VF, the lowest energy that cannot induce VF should be equal to the lowest shock strength that terminates it. Despite this, the DFT is usually slightly above the ULV, with a difference ≥ 10 J traditionally considered clinically significant. This difference can be found in up to 5% of patients, mainly if there is no history of CAD and an absence of documented ventricular arrhythmias. Testing for ULV could reduce the need for repeated fibrillation induction and shocks that testing for DFT requires, but no uniform protocol exists.

B. Basics on ICD and Lead Design

Some significant differences are found in respect to pacemakers. The size of ICD generators is larger than that of pacemakers due to the need for a bigger battery, high-voltage capacitors (for storing energy), the electronic circuitry for computing and control, and a high-voltage transformer (for changing the small battery voltage to high voltage), plus several additional components (antenna for communication, a reed switch that responds to a magnetic field for altering the operation of the ICD, activity sensors, audible alarms, etc). Despite this, modern devices are still small enough to allow for subclavian (pectoral) implantation. Replacement due to battery depletion is expected to occur every 5 to 7 years, but it certainly depends on energy consumption (e.g., percentage of pacing, frequency of shocks).

Leads are usually single-lead, dual-electrode systems; 2 main types of design are used: with integrated bipolar leads, the electrode configuration goes from the distal coil of the high-voltage lead system (anode) to the distal tip electrode (cathode); true bipolar leads are dedicated bipolar sensing leads, with an electrode configuration ring-to-tip. They ensure a better postshock sensing and constitute almost 100% of current implants.

The defibrillation electrodes are large and positioned in a way that increases the flow of current to the myocardium: the proximal coil is usually located at the level of the superior vena cava and the distal coil at the RV apex. There are multiple shock configurations but the most effective seems to be that of a triad, incorporating an "active can": the current vector goes from the RV lead to both the proximal coil in the superior vena cava and the can (generator).

In dual-chamber systems a second sensing lead is implanted at the usual place in the high RA.

C. Defibrillation Threshold Testing

Testing for DFT has been an obligatory step since ICDs were introduced in clinical practice. Ensuring an adequate safety margin, usually considered to be a successful defibrillation at an energy setting < 10 J the maximum output of the ICD, was a means to predict the probability of success in case of future events. This was well accepted in the initial era of ICD implantation, but knowledge gained on how defibrillation occurs, technical developments, and data from clinical trials have brought this practice into question.

As explained, the probability of a successful defibrillation depends on the energy of the shock, with each shock at any given energy level having a probability of success; this also means that, over time, there is always a probability of failure at the same energy level. On the other hand, conditions under which DFT is done at the time of implant are rarely, if ever, encountered in the spontaneous clinical setting. The DFT actually varies widely, even on a day-to-day basis, being influenced by factors like AAD therapy, other cardiac and noncardiac drugs, electrolyte levels, sympathetic tone, and so forth. Testing itself is also not exempt from risk, with significant systemic challenges resulting from the use of general anesthesia and repeated induced fibrillation/defibrillation, most often in the presence of SHD.

With the advent of newer ICD system designs, biphasic waveforms, and the ability to modify the defibrillation configuration using noninvasive means, it has become rare to invasively revise an ICD system due to high DFT. Furthermore, clinical evidence supporting the abandonment of this practice in selected groups of patients has been confirmed by the results of a recent trials; for example, there was no difference in survival between primary prevention patients who had a DFT ≤ 10 J and those who had a DFT > 10 J.

As our knowledge stands now, there is little clinical significance in testing DFT in stable patients receiving an ICD for the purpose of primary prevention. Further information is still necessary on testing for patients receiving CRT devices or having known clinical criteria suggestive of high DFT (lower EF, NIDCM, amiodarone therapy). Testing in patients with implanted ICDs due to secondary prevention indications is still probably warranted.

D. Indications for ICD implantation

ICDs are indicated for the prevention of sudden death due to ventricular tachyarrhythmias in patients with known predisposing factors (primary prevention) or in survivors of malignant tachyarrhythmic events (secondary prevention) (Table 12.1). A special group is that of pediatric and CHD patients (Table 12.2).

Table 12.1. Recommendations for ICD Therapy

Class I

- Survivors of cardiac arrest due to VF or hemodynamically unstable sustained VT after evaluation to define the cause of the event and to exclude any completely reversible causes
- Patients with SHD and spontaneous sustained VT, whether hemodynamically stable or unstable
- Patients with syncope of undetermined origin and clinically relevant, hemodynamically significant sustained VT or VF induced at EPS
- Patients with LV EF < 35% due to prior MI who are at least 40 days post-MI and are in NYHA functional class II or III
- Patients with NIDCM with LV EF ≤ 35% and who are in NYHA functional class II or III
- Patients with LV dysfunction due to prior MI who are at least 40 days post-MI, have a LVEF < 30%, and are in NYHA functional class I
- Patients with NSVT due to prior MI, LVEF < 40%, and inducible VF or sustained VT at EPS

Class IIa

- Reasonable indication in patients with unexplained syncope, significant LV dysfunction, and NIDCM
- Patients with sustained VT and normal or near-normal ventricular function
- Patients with HCM who have one or more major risk factors for SCD
- Reasonable for the prevention of SCD in patients with ARVC/D who have one or more risk factor for SCD
- Reasonable to reduce SCD in patients with LQTS who are experiencing syncope and/or VT while receiving β-blockers

Table 12.1. (*Continued*)

- Reasonable for nonhospitalized patients awaiting cardiac transplantation
- Reasonable for patients with BrS who have had syncope
- Reasonable for patients with BrS who have documented VT that has not resulted in cardiac arrest
- Reasonable for patients with CPVT who have syncope and/or documented sustained VT while receiving β-blockers
- Reasonable for patients with cardiac sarcoidosis, giant-cell myocarditis, and Chagas disease

Class IIb

- ICD may be considered in patients with nonischemic heart disease who have an LVEF ≤ 35% and who are in NYHA functional class I
- May be considered in patients with LQTS and risk factors for SCD
- May be considered in patients with syncope and advanced SHD in whom thorough invasive and noninvasive investigations have failed to define a cause
- May be considered in patients with a familial cardiomyopathy associated with sudden death
- May be considered in patients with LV noncompaction

Class III

- ICD not indicated for patients who do not have a reasonable expectation of survival with an acceptable functional status for at least 1 year, even if they meet the previous ICD implantation criteria
- Not indicated for patients with incessant VT or VF
- Not indicated for patients with significant psychiatric illnesses that may be aggravated by device implantation or that may preclude systematic follow-up
- Not indicated for NYHA class IV patients with drug-refractory CHF who are not candidates for cardiac transplantation or implantation of a CRT device that incorporates both pacing and defibrillation capabilities
- Not indicated for syncope of undetermined cause in a patient without inducible ventricular tachyarrhythmias and without SHD
- Not indicated when VF or VT is amenable to surgical or catheter ablation (e.g., atrial arrhythmias associated with WPW syndrome, RV or LV outflow tract VT, idiopathic VT, or fascicular VT in the absence of SHD)
- Not indicated for patients with ventricular tachyarrhythmias due to a completely reversible disorder in the absence of SHD (e.g., electrolyte imbalance, drugs, or trauma)

Table 12.2. Recommendations for ICD Therapy in Pediatric Patients and Patients with CHD

Class I

- Survivors of cardiac arrest after evaluation to define the cause of the event and to exclude any reversible causes
- Patients with symptomatic sustained VT in association with CHD who have undergone hemodynamic and EP evaluation; catheter ablation or surgical repair may offer possible alternatives in carefully selected patients

Class IIa

- Reasonable indication in patients with CHD with recurrent syncope of undetermined origin in the presence of either ventricular dysfunction or inducible ventricular arrhythmias at EPS

Class IIb

- ICD may be considered for patients with recurrent syncope associated with complex CHD and advanced systemic ventricular dysfunction when thorough invasive and noninvasive investigations have failed to define a cause

Class III

- All class III recommendations found in Table 12.1

II. BASIC PROGRAMMING

All ICDs have standard pacemaker functions, but their primary function is the detection and treatment of potentially fatal ventricular arrhythmias. The type of response (antitachycardia pacing [ATP], shocks) to a particular tachycardia is programmable, with different responses to different tachycardias according to prespecified rate zones (tiered therapy). Another significant feature is the ICD's ability to store large amounts of data, including number and duration of events as well as the actual intracardiac EGMs recorded at the moment of a detected arrhythmia.

A. Sensing

True bipolar sensing allows for a more localized sensing of intrinsic signals because the scanned surface is smaller (while it is broader when integrated bipolar leads are used).

When a depolarization wave passes, it is detected and processed, diagnostic criteria are applied, and, according to the final diagnosis, therapy is withheld or delivered.

The sensing software is more complex than in the case of pacemakers because of the added arrhythmia detection and therapy delivery algorithms and EGM storage capabilities. Another significant difference from pacemakers, which often use fixed-gain sensitivity, is that all ICDs use automatic adjustments of amplifier gain or sensing threshold to ensure appropriate detection of arrhythmias. They do this by increasing sensitivity over time between sensed or paced ventricular events, searching for low-amplitude fibrillatory EGMs that may be missed at lower sensitivities (it also helps prevent sensing of T waves, cross-chamber events, or pacing artifacts).

Marker channels, as in pacemakers, play a very important role in ICDs, identifying what the device "sees," at all times, and on a beat-to-beat basis. They show pace/sense annotations and ICD function annotations (sensing, detection, and therapy).

B. Detection

The detection of an arrhythmia is based on what is sensed by the device, with the confirmation of the sensed rhythm as an arrhythmia based primarily on its measurements of rate and duration. These measurements are carried out in beat-to-beat intervals (milliseconds) and/or beats per minute (bpm) and allow the classification of the sensed rhythm in the zones of VT or VF. These detection zones are programmable in ranges of rates and detection duration (measured in NID or length of time to detect) (Figure 12.2). The detection duration is programmable by beat or interval counters (consecutive or probabilistic), or according to time in seconds. A third programmable zone, the fast VT zone, is optional and programmed via VT (using consecutive counters) or via VF (probabilistic counter).

Having more than one detection zone allows tailoring of therapy to the arrhythmia—for example, ATP for slower VT, cardioversion for fast VT, and shocks for VF.

Noncommitted therapy (reanalysis of the rhythm before therapy is delivered) and dual-chamber sensing help reduce inappropriate detection/therapy. Several other programmable features are also used with the same objective (Figure 12.2):

- **Sudden onset criterion:** Considers the degree of prematurity of the initial beat of the VT with respect to previous ones (usually ≥ 9% decrement); it works very well as a differentiator from sinus tachycardia.
- **Stability:** Refers to the regularity of the tachycardia, usually considered stable if there is < 40 milliseconds in mean CL variability. This criterion is useful to differentiate VT from AF but it lacks reliability in cases of AF with ventricular response > 170 bpm (due to pseudo-regular ventricular rate), or in patients on amiodarone or Ic AADs (because VT could become irregular, slow, or polymorphic).
- **Sustained rate criterion:** Creates a delay of 30 to 120 seconds before initiating therapy if onset or stability criteria were not met but rate was still met

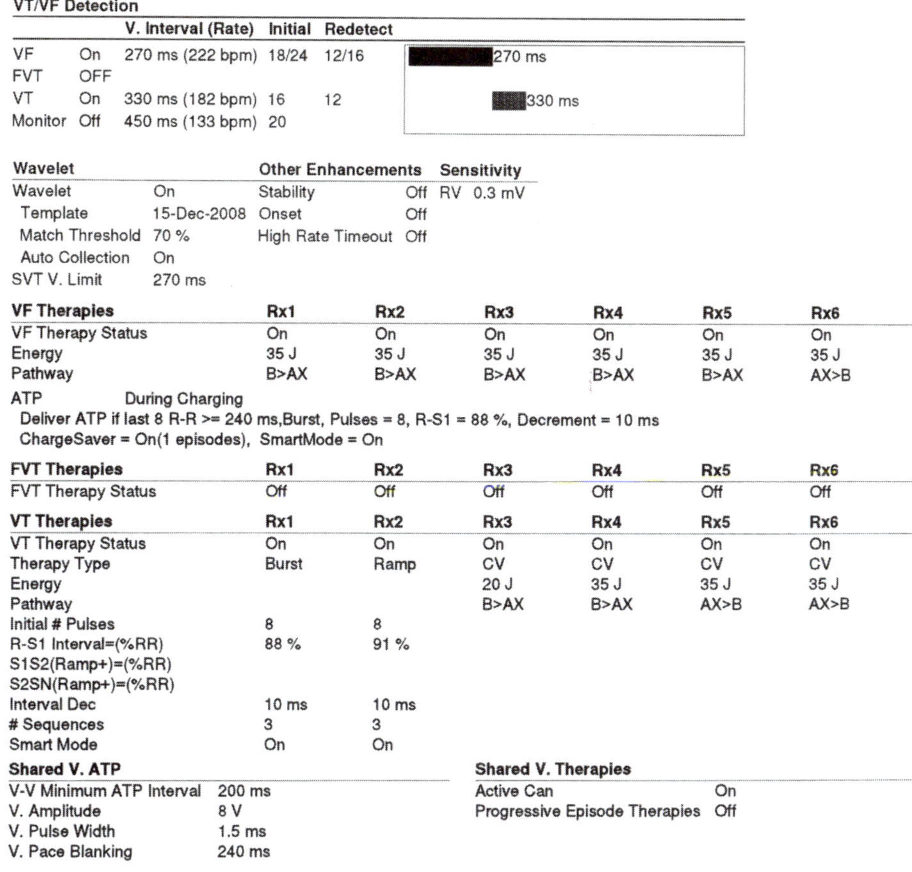

VT/VF Detection

		V. Interval (Rate)	Initial	Redetect
VF	On	270 ms (222 bpm)	18/24	12/16
FVT	OFF			
VT	On	330 ms (182 bpm)	16	12
Monitor	Off	450 ms (133 bpm)	20	

270 ms
330 ms

Wavelet		Other Enhancements		Sensitivity	
Wavelet	On	Stability	Off	RV	0.3 mV
Template	15-Dec-2008	Onset	Off		
Match Threshold	70 %	High Rate Timeout	Off		
Auto Collection	On				
SVT V. Limit	270 ms				

VF Therapies	Rx1	Rx2	Rx3	Rx4	Rx5	Rx6
VF Therapy Status	On	On	On	On	On	On
Energy	35 J	35 J	35 J	35 J	35 J	35 J
Pathway	B>AX	B>AX	B>AX	B>AX	B>AX	AX>B

ATP During Charging
Deliver ATP if last 8 R-R >= 240 ms,Burst, Pulses = 8, R-S1 = 88 %, Decrement = 10 ms
ChargeSaver = On(1 episodes), SmartMode = On

FVT Therapies	Rx1	Rx2	Rx3	Rx4	Rx5	Rx6
FVT Therapy Status	Off	Off	Off	Off	Off	Off

VT Therapies	Rx1	Rx2	Rx3	Rx4	Rx5	Rx6
VT Therapy Status	On	On	On	On	On	On
Therapy Type	Burst	Ramp	CV	CV	CV	CV
Energy			20 J	35 J	35 J	35 J
Pathway			B>AX	B>AX	AX>B	AX>B
Initial # Pulses	8	8				
R-S1 Interval=(%RR)	88 %	91 %				
S1S2(Ramp+)=(%RR)						
S2SN(Ramp+)=(%RR)						
Interval Dec	10 ms	10 ms				
# Sequences	3	3				
Smart Mode	On	On				

Shared V. ATP	
V-V Minimum ATP Interval	200 ms
V. Amplitude	8 V
V. Pulse Width	1.5 ms
V. Pace Blanking	240 ms

Shared V. Therapies	
Active Can	On
Progressive Episode Therapies	Off

Figure 12.2. Example of ICD antitachycardia programming. © Medtronic, Inc. 2010.

at the end of this period. It is very useful as a backup mechanism, especially if the sudden onset criterion is used.

- Rate averaging: Leads to therapy with rates below the programmed VT detection zone if they are close enough to it.
- **Morphology algorithms (e.g., Morphology Discrimination, EGM Width, Wavelet Dynamic Discrimination):** All are based on EGM morphology: a template of the native QRS is compared with the QRS during tachycardia, giving points according to the degree of match or mismatch, and based on the result therapy is provided or withheld. These algorithms should not be used in cases of preexistent BBB because there could be a match between the template and the VT, with therapy withheld; another disadvantage is that they could prevent therapy for narrow complex VT.

- **Enhanced discrimination criteria (e.g., AF Rate Threshold Criterion, SMART, PARAD, PR Logic):** Utilized in dual-chamber systems, these are more sophisticated algorithms that analyze the chamber of origin of the tachycardia, the chamber with the highest rate (ventricular rate faster than atrial rate points towards VT), the chamber of acceleration (e.g., AT begins with a short PP interval followed by a short R-R interval, whereas VT begins with a short R-R interval and a few beats of AV dissociation until 1:1 VA conduction stabilizes), the AV conduction (1:1 conduction, P:R pattern, PR vs RP intervals, PR interval stability, etc), and the response to ATP (e.g., termination of a tachyarrhythmia by a single trial of ATP indicates VT).

Monitoring zones are especially designed to monitor and diagnose presumably slower, more stable ventricular tachyarrhythmias; they are very helpful in patients with unexplained symptoms (palpitations, presyncope/syncope) and to monitor the effect of AADs on the features of the arrhythmia (CL, morphology). Programming of monitoring zones varies from one manufacturer to the other; they are not necessarily passive features as interaction/interference with therapy zones is possible (see following for details).

C. Therapy

1. Antibradycardia pacing

For patients not having a pacing indication this function is a demand mode only (VVI), useful to avoid bradyarrhythmias, primary or secondary to other arrhythmias, ATP, or shocks. In the presence of concomitant pacing indications, pacing in ICDs has similar rationale and use as in pacemakers. Dual-chamber ICDs are also better discriminators between supraventricular and ventricular tachyarrhythmias.

2. Antitachycardia pacing

ATP allows overdrive pacing of a VT, thereby decreasing the need for cardioversion or defibrillation. It has been found to be effective in 89% to 95% of episodes of spontaneous reentrant VT (Figure 12.3). ATP works by pacing the heart slightly faster than the VT rate and is programmable in several ways, all with comparable results: (1) burst pacing at equal intervals; (2) adaptive burst pacing, which paces at a programmable percentage (80%–90%) of the VT CL; and (3) ramp pacing, which allows decrements (usually of 10 milliseconds) in the CL of each consecutive pacing stimulus (it may also add a pulse per sequence).

3. Cardioversion

Cardioversion allows low-energy shocks (≤ 2 J), synchronized to the R wave, in an attempt to stop a reentrant arrhythmia without recurring to high-energy therapy. Cardioversion is as effective as ATP and preferred over it when ATP results in unwanted

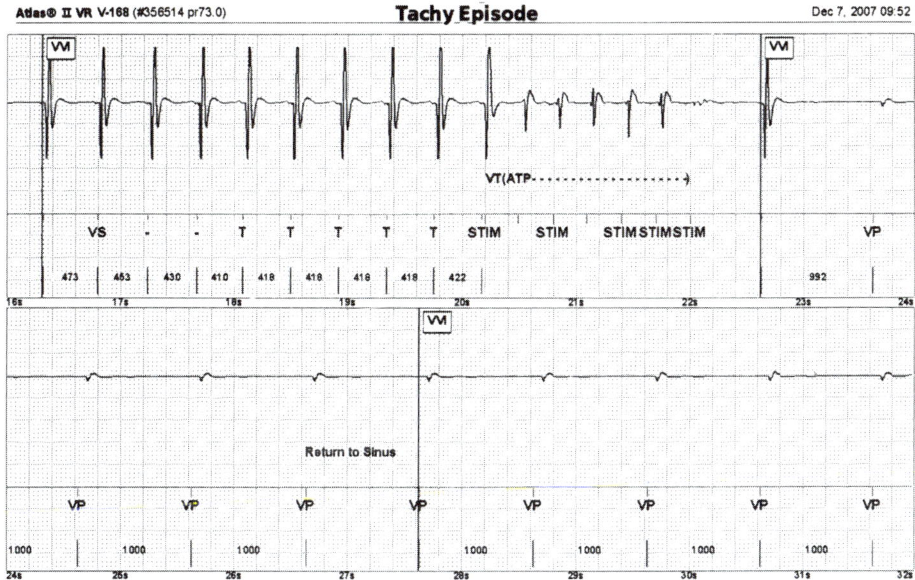

Figure 12.3. ATP successfully stops a VT. After therapy, all detected events are ventricular paced ones (VP).

effects (acceleration of the arrhythmia, syncope due to delayed shock therapy). As it is for ATP, low-energy shocks are preferred to treat hemodynamically stable VT with not very high rates (< 150 bpm) because they may delay a more appropriate therapy in cases of VT with serious clinical repercussions.

4. Defibrillation

High-energy shocks are among the most effective ways for stopping a ventricular arrhythmia (Figure 12.4). The energy delivered varies according to the manufacturer but may be up to 40 J. Modern devices are noncommitted: they must reconfirm the continued presence of the initially detected arrhythmia during charging and/or at the end of charging prior to delivering therapy (this avoids inappropriate shocks for NSVT). Some device models are also capable of delivering ATP during charging.

III. PRINCIPLES OF ICD IMPLANTATION AND FOLLOW-UP

In the absence of a pacing indication, single-chamber devices should be preferred, programmed to avoid ventricular pacing (e.g., VVI at 40 bpm), as this could increase morbidity and mortality. Dual-chamber ICDs have better VT detection capabilities, but they are not indicated in patients without a pacing indication or recurrent supraventricular arrhythmias.

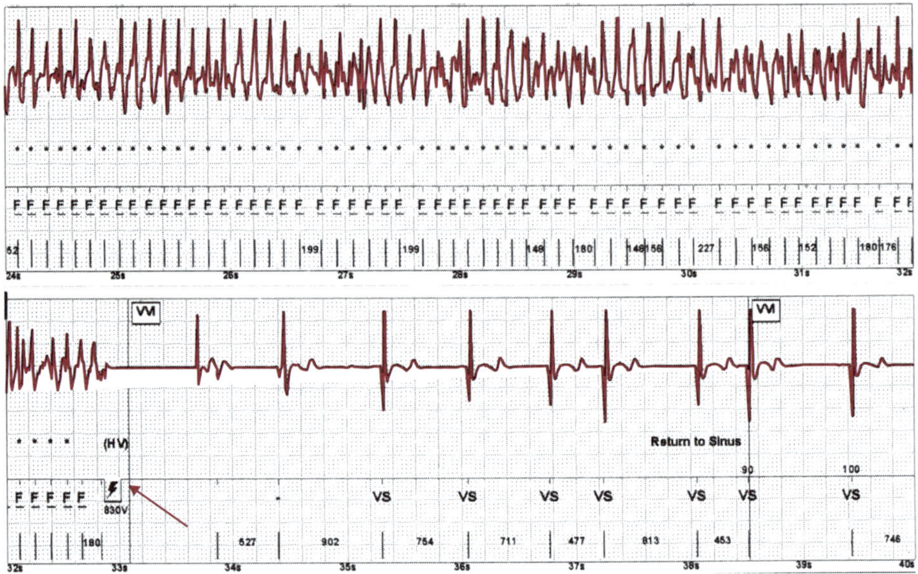

Figure 12.4. Once VF detection criteria are met, a shock (arrow) successfully stops the arrhythmia.

ICD implantation follows the same principles as a pacemaker (see Chapter 11) in terms of preoperatory care, anticoagulation therapy, surgical approach, and post-procedure care. The surgical risks are also approximately the same. Specific for ICDs is the need for DFT testing in selected patients (see previous for rationale).

At implantation R-wave sensing should ideally be > 8 mV (though ≥ 5 mV is accepted), with the programmed sensitivity commonly set to ≤ 0.3 mV; the pacing threshold should be < 1 V for a PW of 0.4 milliseconds (threshold < 1.5 V is accepted). If DFT is to be done, a waiting period of 3 to 5 minutes between shocks is advised.

As with pacemakers, ICDs should be followed up in dedicated clinics by personnel experienced in dealing with their intricacies in programming and troubleshooting (Table 12.3).

Periodic visits should be scheduled every 3 to 6 months, with due care paid to indicators of battery depletion (battery voltage, impedance, charge time).

Patients should be properly orientated on what to expect and what to do in the event of a shock, which could happen in up to one-half of all ICD recipients in the first year postimplant. Apart from the sensation perceived, from tolerable to severe according to the patient's level of tolerance, a shock can also trigger feelings of fear and anxiety. Appropriate counseling should include the recommendation of seeking immediate medical help in cases of significant symptoms (chest pain, dyspnea, presyncope/syncope) or recurrent shocks (2 or more in a 24-hour period). Patients who receive their first shock or who have isolated, noncomplicated episodes should contact their physicians to arrange adequately timed follow-ups.

Table 12.3. Steps in Reviewing a Patient Post-ICD Implantation

1. Focused questioning (e.g., occurrence of shocks)
2. Focused physical examination (scar, generator pocket)
3. Full medication review
4. Capacitor charge time, battery voltage and impedance, lead impedance (including shocking impedance)
5. Atrial/ventricular capture and sensing thresholds
6. Review of programmed pacing parameters
7. Review of programmed detection criteria and therapy for VT/VF
8. Review of recorded episodes of arrhythmia detection and delivered therapy
9. Review of alerts
10. Optimizing settings if necessary/possible
11. Troubleshooting if needed

Because adequate ICD functioning depends on proper sensing, avoidance of sources of electromagnetic interference should be carefully exercised: for example, cell phones held no less than 15 cm from the ICD, no lingering around electromagnetic security systems, avoidance of MRI testing, and so forth.

IV. ICD COMPLICATIONS AND TROUBLESHOOTING

Abnormalities in capture and sensing, as well as complications not related to system malfunction, occur in ICDs just as they do in pacemakers and should be managed in a similar fashion (see Chapter 11). Complications present only with ICDs are obviously related to their antitachycardia function and include problems with proper detection of arrhythmias and the delivery of therapy.

A. Oversensing

Signals, physiological or not, may be oversensed and interpreted as arrhythmias by the ICD (Table 12.4). Oversensing is probably infrequent and most cases of overdetection, leading to inappropriate therapy, are due to SVT.

Oversensing of intracardiac signals leads to multiple counting; this can then be detected as VT or VF by the ICD. Double counting of R waves happens if the duration of the sensing EGM exceeds the ventricular RP of 120 to 140 milliseconds; when it occurs on the atrial channel of a dual-chamber system it does not lead to inappropriate detection of VT if the ventricular rate is in the sinus zone, but it may confound SVT/VT discrimination if criteria for VT are fulfilled. T-wave oversensing in the presence of small R waves (< 3 mV) may be corrected by using programming features (like "decay delay"), but most often invasive lead revision/repositioning or adding a separate sense/pace lead is required (Figure 12.5).

Table 12.4. Oversensing in ICDs and Troubleshooting Options

Causes	Characteristics	Troubleshooting Options
Intracardiac signals (P, R, or T waves)	Intracardiac EGM: characteristic alternation of R-R intervals and EGM morphology separated by isoelectric baselines May be detected as VT or VF	P-wave oversensing: forced atrial pacing or lead repositioning R-wave oversensing: increasing the ventricular blanking period and/or decreasing sensitivity T-wave oversensing: increasing postpacing ventricular blanking period (for postpacing T oversensing), decreasing sensitivity, using specific programming features or lead repositioning (if spontaneous T-wave oversensing)
Extracardiac signals (EMI, myopotentials)	Normal lead appearance on the chest x-ray Intracardiac EGM: replacement of the isoelectric baseline with high-frequency noise May be associated with postural changes (myopotentials) R-R intervals often sensed close to the RP of the ICD (120–140 ms) Almost always detected as VF	Avoidance of identifiable causes of EMI Decreasing ventricular sensitivity or lead repositioning (myopotentials)
Lead wire fracture Lead partially inserted	Abnormal lead appearance on the chest x-ray May be associated with postural changes Abnormal lead tests	Perform invasive test and treat accordingly

Forced atrial pacing in cases of P-wave oversensing shortens the ventricular CL to prevent ventricular sensitivity from reaching its minimal value and introduces cross-chamber ventricular blanking after each atrial event; in all instances dislodgement of the ventricular lead should be excluded.

Electromagnetic interference due to surgical electrocautery has been a source of inappropriate shocks to both patients and medical personnel when therapy has not been adequately suspended before surgery. Myopotential oversensing, often diaphragmatic, may lead to syncope in pacemaker-dependent patients (due to pacing inhibition), followed by inappropriate shock (exception to the rule that syncope followed by shock indicates appropriate shock); repositioning the lead should solve the problem.

In all situations where decreasing ventricular sensitivity has been chosen as a therapeutic option, the operator must make sure that sensing of VT/VF is reliable at the reduced level of sensitivity.

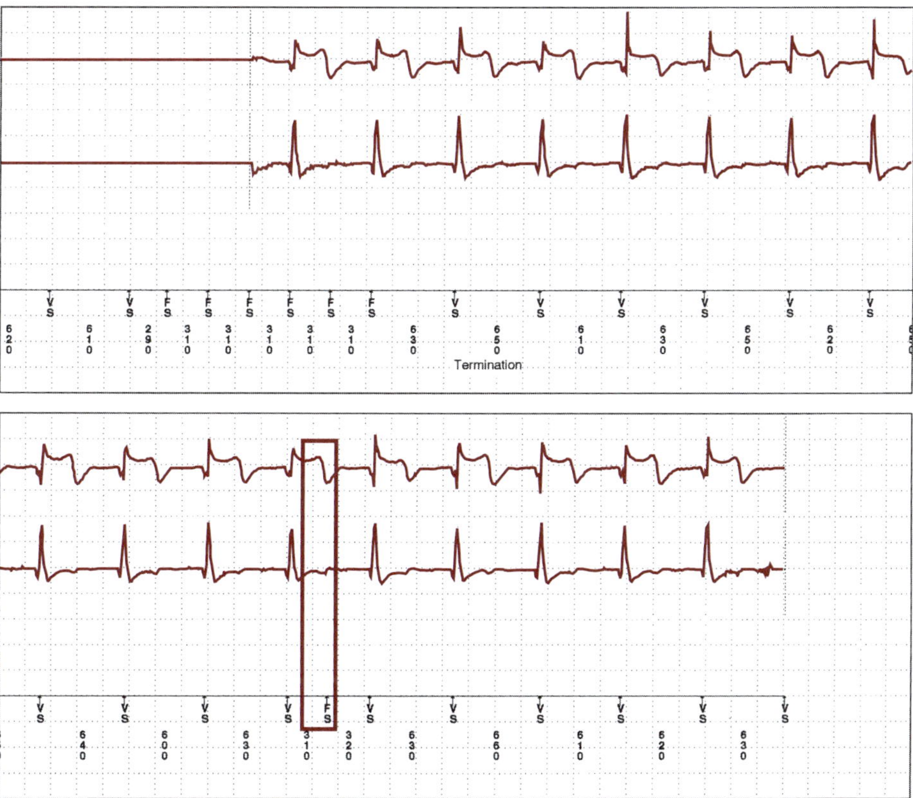

Figure 12.5. Intermittent T-wave oversensing.

B. Undersensing

Undersensing leads to failure or delay to deliver therapy, which can be fatal (Table 12.5).

If the amplitude of native R waves is ≥ 5 mV it is rare to find significant undersensing. Variations in amplitude of the intracardiac EGM leading to undersensing may also cause detection in the wrong zone, initiation of an inappropriate ATP CL, or inappropriate redetection of "sinus rhythm." Dropout could be more evident after an ineffective shock due to changes in R waves; this is probably lead-dependent,

Table 12.5. Undersensing in ICDs and Troubleshooting Options

Causes	Characteristics	Troubleshooting Options
Programming issues (sensitivity setting, inappropriate rate cutoff)	Normal lead appearance on the chest x-ray Reprogramming normalizes sensing	Reprogram settings accordingly
Postshock changes (dropout, tissue changes)	Normal lead appearance on the chest x-ray Intracardiac EGM: variations in amplitude during VT	Shortening the time or numbers required for arrhythmia detection (TDI or FDI) Increasing maximum sensitivity Making therapy committed Accelerating the rate at which the sensing amplifier increases its gain
Lead wire fracture Lead partially inserted Generator problems	Lead appearance may be abnormal on the chest x-ray May be associated with postural changes Abnormal lead and/or generator tests	Perform invasive test and treat accordingly
Miscellaneous: drug or electrolyte effects (mostly hyperkalemia), ischemia	Normal lead appearance on the chest x-ray Normal lead tests Suggestive clinical history and problem-oriented testing	Remove offending agents Manage clinical condition

more frequently with integrated RV defibrillation leads. Atrial undersensing during exercise may provoke inappropriate ICD discharge. In all cases where there is a failure to deliver therapy, the operator must verify that the therapy delivery has been appropriately activated.

C. Delivery of Inappropriate Therapy

Most cases of inappropriate therapy are due to oversensing of SVTs and NSVT, electrical noise (e.g., electromagnetic interference), and ICD malfunction (e.g., lead fracture) (Figure 12.6). Inappropriate shocks occur in 20% to 25% of ICD patients; when multiple, their effect on the patient is very traumatic, with significant pain, stress, and anxiety.

Reducing the risk of inappropriate shocks is achieved by using noncommitted therapy, dual-chamber sensing, programming of SVT discriminators, optimization of sensing, AAD therapy, and catheter ablation of SVTs.

Inappropriate therapy due to interference from monitor-only zones is rare, but occasionally the programming features used to define a monitoring zone, which by principle is to be a passive zone, may interfere with the criteria utilized in therapy zones and can lead to inappropriate detection and therapy. Because the programming varies among the different ICD manufactures, an in-depth technical knowledge is necessary to deal with these uncommon events. Involved mechanisms include combined, cumulative, and shared counters, as well as maximum time to therapy timers. Newer ICD models use varied methods to avoid this problem, but older models may still deliver therapy in monitoring-only zones. Therapeutic options are found in Table 12.6.

Newer models from Medtronic have true monitoring zones that do not activate the combined count criterion. No shocks in monitoring zones have been reported with ICDs from Boston Scientific (formerly Guidant) or ELA.

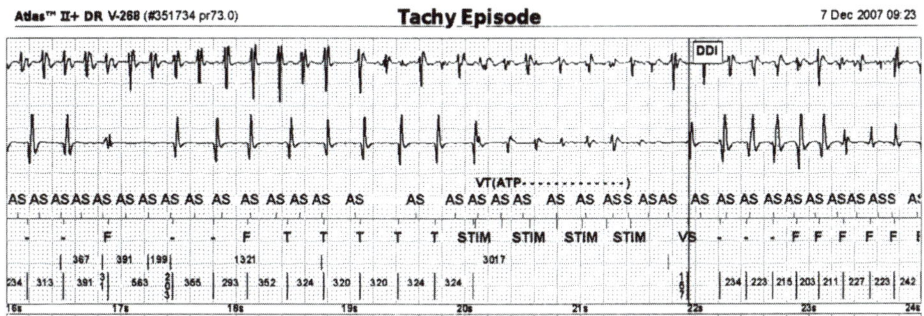

Figure 12.6. A rapidly conducted atrial tachyarrhythmia leads to inappropriate detection of VT and subsequent ATP, obviously nonsuccessful.

Table 12.6. Interference due to Monitor-only Zones

Manufacturer and Models	Mechanism Involved	Troubleshooting Options
Medtronic: devices older than Marquis 7274 and 7230	Combined count criterion: because it is activated whenever a VT zone is "on" (even if therapies are programmed to "off")	Program the pseudo-monitoring zone "off" Increase the VF detection threshold
St. Jude Medical: devices older than series II	Slowing of ongoing arrhythmia, going from higher zone (like VF or VT-2) to monitoring zone: if post-VF or VT-2 detection interval/rate is programmed within the monitoring zone Timing out of an MTF timer Binning faster intervals	Program the post-VF or VT-2 detection interval/rate "same as VF" or "same as VT-2" Consider a high VF (and VT-2) threshold to minimize risk of binning a fast interval Program VT therapy timeout "off" If VT therapy timeout is "on," program the longest timeout zone Program sinus redetection at 3 instead of 5 bins to increase the likelihood of earlier closure of the monitoring zone and reset faster bins Increase the number of intervals required to trigger the monitoring zone

D. High Defibrillation Threshold

The use of biphasic waveforms, active cans, and left pectoral implants has resulted in lower DFT compared to older approaches, but still at the time of DFT testing or when a shock fails to convert a ventricular tachyarrhythmia the physician may find a high DFT. It is estimated that high DFT occurs in 5% to 6% of patients (3% in a study including high-output devices) and that a high percentage of these patients die due to SCD, probably secondary to inadequate defibrillation energy. Several strategies have been devised to solve this problem (Table 12.7).

In all instances other possible complications must be ruled out: pneumothorax, medications that increase DFT (mexiletine, carvedilol, sildenafil, amiodarone), and temporal situations (anaesthesia, long operatory time). Sotalol, known to decrease the DFT, may be tried.

Table 12.7. Therapeutic Strategies in Patients with High DFT

- Use of high-output device
- Repositioning of the distal coil (more apical) or the RV lead (RV outflow tract)
- Use of subcutaneous array
- Repositioning of the proximal electrode (left subclavian, brachiocephalic, or azygos vein)
- Removal of the superior vena cava coil, leading to unipolar configuration (attractive in patients with impedance < 40 Ω)
- Placing an auxiliary lead in the coronary sinus, the superior vena cava, or a subclavian vein axillary position
- Changing polarity, using the distal coil as the anode (distal coil as the cathode is the nominal configuration)
- Modifying tilt of the waveform

E. Acceleration of Ventricular Tachycardia

Acceleration of VT is a ≥ 10% decrease in the CL of a monomorphic VT after delivery of ATP therapy; another definition includes the ATP-induced transformation of a stable VT into polymorphic VT or VF. Acceleration of VT is most likely to occur with Ramp pacing than with Burst, even though they could have similar effect in cases of slow VT (Figure 12.7). ATP therapy should be reconsidered whenever hemodynamically significant fast VT occurs; if conversion to polymorphic VT/VF happens, then all ATP should be discontinued and initial shocks programmed.

F. Multiple Shocks

Multiple shocks are not necessarily an ICD complication because recurrent episodes of VT (electrical storm), properly detected and successfully treated, may occur. Multiple shocks become an ICD-related problem when inappropriate detection occurs, or when appropriate detection leads to unsuccessful shocks. Oversensing should be managed as previously explained. Failure of 2 or more maximum-output shocks, with appropriate detection of VT/VF, should not occur, unless the safety margin is not adequate; if it happens, patient-related (pneumothorax, pleural effusion) or system-related reasons should be sought.

G. Manufacturer's Advisory/Recall

Device advisories/recalls usually portend a more complicated situation that that of pacemakers. Apart from possible complications in regard to its pacing function, fail-

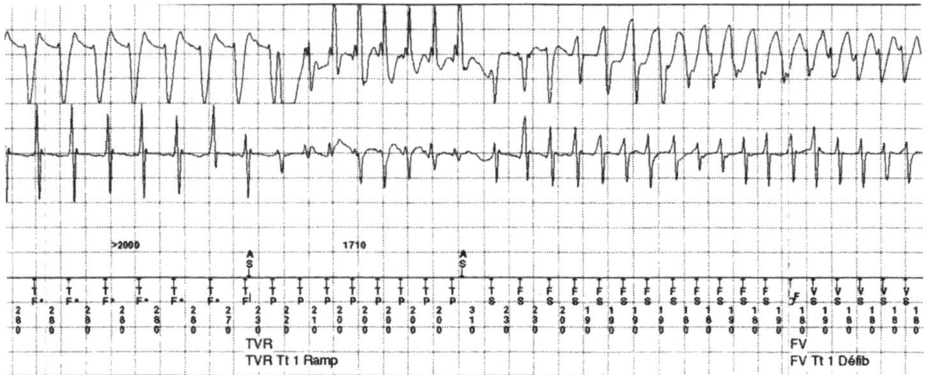

Figure 12.7. Acceleration of VT due to Ramp ATP.

ure to appropriately sense, detect, or treat a ventricular arrhythmia defeats the very purpose an ICD is implanted for, which is to save lives. Again, specific recommendations on how to deal with these problems usually come from the manufacturers and though occasionally a noninvasive approach is suggested (software/programming changes, increased follow-up), in many instances a system extraction/replacement is needed.

V. ADJUNCTIVE THERAPY WITH ICDs

A. Antiaarhymic Drugs

Antiarrhythmics are prescribed in up to 70% of patients with ICDs to reduce the frequency of shocks; to avoid high-energy shocks, by terminating ventricular arrhythmias together with ATP or low-energy cardioversion; to reduce the rate of VT so it becomes more tolerable; and to treat concomitant supraventricular tachyarrhythmias. The therapy usually consists of amiodarone or sotalol. Dofetilide, azimilide, and dronedarone are possible options.

Because of the AAD effect on the tachycardia CL and QRS morphology, once a drug is initiated in a patient already carrying an ICD it is advisable to update the arrhythmia detection and therapy delivery criteria, and even to test the DFT if deemed necessary. This seems more appropriate in patients taking amiodarone, which has been associated with an increase in DFT. Most AADs have no effect on, or are even associated with a decrease in, the DFT.

B. Catheter Ablation

If AAD therapy fails to decrease the frequency of shocks, due to recurrent episodes of VT, percutaneous catheter ablation of the arrhythmia circuit may be indicated. In patients suffering from symptomatic recurrent SVTs or having inappropriate shocks, EPS and ablation of the associated arrhythmia are also indicated.

SUGGESTED READINGS

Epstein AE, Dimarco JP, Ellenbogen KA, et al. ACC/AHA/HRS 2008 guidelines for device-based therapy of cardiac rhythm abnormalities. *J Am Coll Cardiol.* 2008;51:2085-2105.

Mansour F, Khairy P. ICD monitoring zones: intricacies, pitfalls, and programming tips. *J Cardiovasc Electrophysiol.* 2008;19(5):568-574.

Swerdlow CD, Friedman PA. Advanced ICD troubleshooting: part I. *PACE.* 2005;28(12):1322-1346.

Swerdlow CD, Friedman PA. Advanced ICD troubleshooting: part II. PACE 2006;29(1):70-96.

Chapter 13

Selected Techniques and Interventions

Miguel A. Barrero Garcia and Paul Khairy

I. ELECTROPHYSIOLOGICAL STUDY

A. General Principles

The EPS is a catheter-based test of the electrical conducting system of the heart. It is indicated to identify the cause of a tachyarrhythmia or bradyarrhythmia, locate its origin, assess its properties, and/or determine the propensity to develop potentially dangerous arrhythmias. An EPS is also critical in diagnosing the arrhythmia substrate for subsequent catheter ablation and assessing therapeutic efficacy. Common terms and concepts relevant to an EPS include:

- **AH interval:** Time between the onset of the first rapid atrial deflection and the His bundle deflection, which reflects AV nodal conduction time.
- **HV interval:** Time between the onset of the proximal His bundle recording and the earliest ventricular activation, reflecting conduction over the His-Purkinje system.
- **Decremental conduction:** Delay of propagation of an impulse resulting from progressive shortening of the preceding coupling interval. It is a property of the AVN.
- **Effective RP:** The longest coupling interval between the basic drive and a premature impulse that fails to be conducted over the tissue; for example, the AVN effective RP is the longest A_1 to A_2 interval measured at the level of the His bundle that fails to propagate to the His bundle.
- **Functional RP:** The shortest coupling interval between 2 consecutively conducted impulses, as measured by points distal to that tissue; for example, the functional RP of the AVN is the shortest H_1 to H_2 interval that can be elicited in response to any A_1 to A_2 interval.
- **Relative RP:** The longest interval between the basic drive and a premature impulse that results in latency between the premature impulse and drive; for example, the relative RP of the atrium is the longest S_1 to S_2 interval that produces a delay in S_2 to A_2 conduction (i.e., $S_2–A_2 > S_1–A_1$).
- **Patterns of retrograde atrial activation:** The pattern of retrograde atrial activation can help distinguish between normal conduction through the AVN (concentric) or eccentric conduction via an accessory pathway. With VA activation through an AV nodal fast pathway, the earliest atrial activation is recorded near the His-bundle catheter. Coronary sinus activation proceeds in a proximal-to-distal sequence. In the presence of a retrograde-conducting accessory pathway remote from the His-bundle position, the activation sequence will differ from the concentric pattern, with the earliest site of atrial activation reflecting the site of the accessory pathway.
- **Gap phenomenon:** A conduction gap is a zone defined by the failure of premature impulses to conduct, whereas impulses of greater and lesser prematurity successfully propagate. It is thought to reflect differences in RPs

between proximal and distal sites. For example, a gap in AV conduction may be elicited if the proximal portion of the AVN has a shorter effective RP than the distal portion. As the coupling interval is decreased, the impulse will first block in the distal portion of the AVN due to its longer RP. However, as the coupling interval is further shortened, the impulse may be slowed in the proximal portion of the AVN (i.e., relative RP), allowing the distal portion to recover excitability and resume conduction.

B. Technique

The basic EPS consists of placing 2 to 4 electrode-tip catheters via femoral venous access, recording conduction intervals, and performing programmed atrial and ventricular stimulation with incremental pacing and single or multiple premature beats during sinus or paced rhythms. As shown in Figure 13.1, catheters are generally positioned to record signals from the high RA, bundle of His, and RV apex. Left-sided signals may be accessed by placing a multipolar catheter within the coronary sinus to record EGMs along the left AV groove. Additional catheters may be introduced as required.

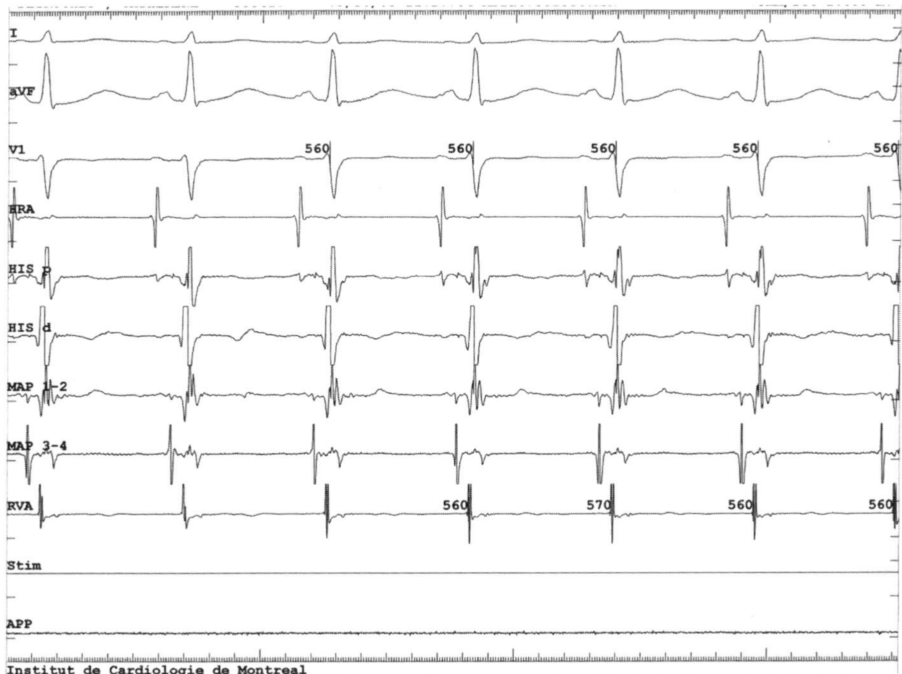

Figure 13.1. Typical intracardiac recordings during EPS. The first 3 tracings represent surface ECG leads I, aVF, and VI. The following 6 tracings are intracardiac EGMs recorded from the high RA (HRA), proximal (HIS p) and distal (HIS d) His bundle, distal (MAP 1–2) and proximal (MAP 3–4) roving mapping catheter, and RV apex (RVA).

Table 13.1. Normal Values during EPS

AH interval: 55–125 ms
His duration: 15–25 ms
HV duration: 35–55 ms
QRS duration: < 100 ms
Atrial effective RP: 180–330 ms
AVN effective RP: 250–400 ms (antegrade conduction)
AVN functional RP: 330–550 ms
Ventricular effective RP: 180–290 ms

A basic EPS includes the following steps: measurement of basic intervals, incremental pacing, atrial and ventricular extra-stimulus testing, measurement of RPs, and specific pacing protocols tailored to the suspected rhythm disturbance, including assessment of SAN function.

Commonly accepted normal ranges are summarized in Table 13.1. Parameters for SAN function are discussed in Chapter 3.

II. TEMPORARY PACING

A. General Principles

Temporary pacing is generally used to provide transient rate support from an external pulse generator via an electrode, or electrodes, for a brief period of time. It is employed either as a bridge to permanent pacemaker implantation or while awaiting resolution of the bradyarrhythmia, either spontaneously or following correction of a reversible cause. Various approaches are possible and may be tailored to the clinical scenario, including transvenous, transcutaneous, transthoracic, epicardial, and transesophageal approaches.

B. Technique

In the acute cardiothoracic postoperative setting, temporary pacing is most commonly achieved by epicardial wires. Otherwise, the transvenous approach is most frequently used. Intravascular access is obtained from internal jugular, subclavian, or femoral veins. From these sites, a pacing catheter is typically positioned at the RV apex. In patients with severe SND but intact AVN function, the lead may alternatively be placed in the RA. Occasionally, 2 temporary wires may be inserted for synchronous AV pacing. The operator should ensure optimal positioning, so as to avoid complications (see Chapter 11).

Once the catheter is deemed stable, sensing and pacing thresholds should be verified. Ideally, sensing thresholds in the RA and RV should exceed 1 mV and 6 mV, respectively. Sensing is characteristically programmed to a value that is 25% to 50%

lower than this threshold. To verify the sensing threshold, the pacing rate should be set below the intrinsic HR. The sensing value is progressively increased until the native complex is no longer perceptible. The sensing threshold is defined as the highest value that detects the native complex.

In both chambers, optimal pacing thresholds are < 1 V. To allow a sufficient margin of safety, termporary pacemakers are generally programmed to values 3 to 5 times higher than the pacing threshold. To assess the pacing threshold, the pacing rate must exceed the intrinsic HR. Pacing is initiated at high outputs (e.g., 10 V), and the voltage is gradually decreased until loss of capture. The pacing threshold is defined as the lowest value that entails depolarization of the underlying tissue.

A chest x-ray or fluoroscopic imaging postprocedure is recommended to assess proper placement and to exclude potential complications (e.g., pneumothorax).

Transcutaneous pacing generally requires higher levels of energy to capture underlying myocardium and may, therefore, be associated with substantial discomfort from skeletal muscle contractions. In an attempt to minimize distress, electrodes should be positioned in areas that do not directly overlie skeletal muscle, such as the midline chest and below the left scapula. The lowest effective current should be used and sedation considered if these measures are inadequate. To program transcutaneous pacing, the desired HR is set and the current is progressively increased as tolerated until capture is achieved. Typically, a PW between 20 and 40 milliseconds with a 140- to 200-mA current is required. For effective capture, it is essential that the anterior chest electrode be of negative polarity. In the absence of an intrinsic heart rhythm, a palpable pulse indicates successful capture.

C. Complications

The most common complications are a consequence of vascular access and manipulation of catheters within the heart. These include inguinal hematoma, fistula, or aneurysm, particularly in the setting of arterial puncture; venous thrombosis or thrombophlebitis; penetration/perforation of the heart, with subsequent pericardial effusion or tamponade (notably, these complications may become apparent upon retrieval of a temporary wire that entirely traversed the myocardium yet impeded flow into the pericardial sac); bacteremia; thromboembolic events, such as pulmonary emboli, acute MI, or stroke; ventricular tachyarrhythmias; and hemothorax or pneumothorax, particularly in the setting of subclavian venous access.

III. ELECTRIC CARDIOVERSION

A. General Principles

ECV is a brief procedure whereby an electrical shock is delivered to the heart to convert an abnormal heart rhythm back to normal. It temporarily depolarizes most cardiac cells, interrupting a tachyarrhythmia circuit or disorganized rhythm if present and allowing the SAN to resume its pacemaker activity, if functional. In patients with digitalis-induced tachyarrhythmias, ECV may be ineffective, since a homogenous depolarization state already exists. Indeed, it may be proarrhythmic in this context, inducing postshock VT or VF.

The discharge of current should be synchronized to the early portion of the QRS complex, either R or S wave, to minimize the probability that VF will be inadvertently induced. If ECV is indicated but synchronization is not reliable, a high-energy unsynchronized shock is preferred. Modern external defibrillators deliver current with a biphasic waveform, which requires less energy than older monophasic systems.

Indications for ECV may be urgent, as in VT or SVT with hemodynamic compromise, or elective, as in well-tolerated AF, AFL, or AT. ECV is contraindicated for tachyarrhythmias associated with digitalis toxicity and is ineffective for automatic arrhythmias (e.g., sinus tachycardia, ectopic AT, and MAT).

B. Technique

Clinic management should adhere to BLS/ACLS guidelines, with IV access, airway management equipment, and availability of a crash cart.

Short-acting agents (e.g., midazolam) are commonly employed as first-line sedatives and are usually associated with an analgesic opioid such as fentanyl. Antagonist agents such as flumazenil and naloxone should be readily available.

Antero-apical placement of paddles is most common, with one placed to the right of the sternal border at the level of the second/third intercostal space and the second overlying the ventricular apex, typically inferior and lateral to the nipple. In the presence of a pacemaker or ICD, paddles should be placed at least 10 cm away from the device. In the presence of an implantable cardiac rhythm device, or if cardioversion is unsuccessful, an anteroposterior position may be preferable. Such an approach may avoid direct passage of current through the device.

The synchronization mode should be "on." Energy requirements may vary according to the type of arrhythmia. Monomorphic VT may be treated initially with 100 J, with higher output if necessary. For AF, it is common to begin with 100 to 200 J, with up titration if required. For AFL and SVT, 50 to100 J usually suffice.

Vital signs and neurologic status should be carefully monitored until the patient is fully awake. Most may be discharged safely 4 to 6 hours postprocedure, if no other issues require attention. Pacemaker and ICD recipients should have their devices checked prior to discharge, as ECV may provoke device reset, mode reversions, and/or reprogramming from bipolar to unipolar pacing.

C. Complications

Complications are uncommon, but include:

- **Skin "burns":** Most skin reactions may be classified as "first-degree" skin burns and involve redness and edema. They may appear immediately after the shock or develop slowly over a few hours. Redness and discomfort usually subside over 48 to 72 hours. Higher-degree burns are possible, especially in patients with fragile or sensitive skin.
- **Cardiovascular:** Myocardial and epicardial injury may result in arterial hypotension or pulmonary edema. Postcardioversion dysrrhythmias such as bradycardia, asystole, VT, or VF may occur. Stroke is a feared complication, particularly in patients with AF or AFL. Risk should be minimized by adequate precardioversion anticoagulation and/or ruling out the presence of an intracardiac thrombus by transesophageal echocardiography prior to cardioversion.
- **Sedation-related:** Complications include hypoxia/hypoventilation.

IV. EXTERNAL DEFIBRILLATION

A. General Principles

Defibrillation refers to an unsynchronized discharge of electrical energy to interrupt a life-threatening arrhythmia and/or restart the heart rhythm. Many different devices are capable of delivering external shocks, some with built-in software that allows automated recognition of the rhythm disorder and, if indicated, automatic delivery of shocks (automated external defibrillators). In the event of sudden death/cardiac arrest due to VF or VT, early defibrillation is essential as the success rate decreases by 10% per minute following the onset of VF.

External defibrillation is, therefore, indicated to treat VT or VF in the setting of an unresponsive patient *without* a pulse. It should not be used in awake, responsive patients, or in the presence of a pulse.

B. Technique

Until defibrillation becomes available, BLS/ACLS protocols should be initiated. The defibrillator should be powered on and the "unsynchronized/defibrillation" mode selected. Conductive material (gels or pads) should be applied prior to paddle placement, and paddles should be applied as in ECV.

The rhythm should be assessed to confirm VF/VT. If a flat line is present, the gain should be increased to rule out fine VF. If the flat line remains (and technical issues involving the monitor, connections, and patient have been ruled out), the paddles should be rotated 90° and the rhythm reassessed.

The energy level should be selected, usually beginning with 200 J, and the defibrillator charged. "Charge" buttons may be located on the paddles and/or on the central machine. The shock is delivered by simultaneously pressing the discharge buttons on each paddle (or on the monitor for electrode pads), after ensuring that equipment and staff are not in contact with the patient ("all clear").

Cardiopulmonary resuscitation should be pursued while reassessing the patient, with further management tailored according to responses.

C. Complications

The patient may be subject to complications similar to cardioversion. The presence of liquid (body fluids, medications, IV fluids) may cause electrical arching thermal burns to the skin and soft tissue and produce ineffective defibrillation by allowing the current to course superficially, rather than transthoracic. All body fluids or liquids should be wiped away from the skin prior to attempting defibrillation.

Postdefibrillation cardiac dysrhythmias are more common following prolonged VF and high-energy shocks. Early defibrillation at recommended energy levels minimizes this complication.

Electrical injuries to health care providers can result if participants remain in contact with the patient during delivery of a shock. Fires may result from sparks in the presence of nitroglycerin patches or ointment, flammable gasses, or an oxygen-rich environment.

V. CAROTID SINUS MASSAGE

A. General Principles

The carotid sinus is a slightly enlarged portion of the carotid artery where it divides into internal and external carotid arteries. It contains numerous baroreceptors, which function as a "sampling area" for several homeostatic mechanisms that maintain BP control. When hypersensitive, its mechanical deformation results in an exaggerated response characterized by bradycardia and/or vasodilation and, occasionally, hypotension, presyncope, or syncope. Massaging the carotid sinus may help clarify tachyarrhythmia mechanisms, by virtue of slowing AV conduction, unmasking flutter waves, confirming focal atrial tachyarrhythmias, or interrupting reentrant circuits that involve the AVN.

A carotid sinus massage (CSM) may be indicated in the following situations: for the evaluation of suspected cardio-inhibitory syncope, as a diagnostic/therapeutic approach for certain types of supraventricular arrhythmias, and to differentiate SVT with aberrant conduction from VT.

Contraindications include known previous adverse reaction, acute MI, cerebrovascular accident within the last 3 months, history of VT or VF, and carotid bruit or known or suspected carotid artery occlusion.

B. Technique

Medical personnel must continuously monitor the ECG, BP, and HR and have a crash cart at hand. They should also obtain baseline vital signs.

The CSM should be performed with the patient in a supine position, the neck fully extended, and the head turned away from the side being massaged. The carotid pulse should be palpated, midway between the angle of the mandible and superior border of the thyroid cartilage. It should then be massaged for 5 seconds, with the procedure repeated on the contralateral side. If negative, repeating the test with a 70° tilt position increases the sensitivity of the technique by an additional 30%.

C. Types of Response to CSM

1. Normal

A normal response includes transient decrease in sinus rate and slowing of AV conduction.

2. Abnormal response (carotid sinus hypersensitivity)

Abnormal response includes:

- **Cardioinhibitory:** Ventricular asystole for more than 3 seconds, due to either sinus arrest or SA exit block. This response is seen in 70% to 75% of positive cases. AV block may be observed but is far less common.
- **Vasodepressor:** Decrease in systolic BP ≥ 50 mm Hg without an accompanying decrease in HR; or a reduction in systolic BP > 30 mm Hg with clinical symptoms. This type of response occurs in 5% to 10% of cases.
- **Mixed:** Documented in 20% to 25% of cases.

The clinical approach should be guided by the history. An abnormal response in an otherwise asymptomatic individual is not an indication for therapy.

In symptomatic patients with a positive response, predisposing pharmacological agents should be discontinued if clinically justified (e.g., alfa-methyldopa, β-blockers, clonidine, digoxin, diltiazem). A permanent pacemaker implant may be considered.

D. Complications

Asystole should be treated accordingly, although it is usually self-terminating within seconds; the use of a precordial thump and IV atropine may be effective.

Most neurologic sequelae (e.g., paresthesia, visual disturbance, unilateral weakness) resolve within 24 hours. Persistent neurological sequelae are uncommon (0.1%). Treat mainly with aspirin and observation; further management according to evolution.

VI. TILT TABLE TESTING

A. General Principles

TTT is a technique used to investigate syncope, particularly when it is suspected to be neurally mediated. A head-upright position permits the assessment of compensatory processes related to positional change, mediated through the autonomic nervous system. Although the specificity of TTT is high (80%–90% with a testing angle between 60° and 80°), its sensitivity is modest (40%–60%). Provocative drugs, such as isoproterenol, increase the sensitivity of the test, with a mild reduction in specificity.

Indications for TTT include:

1. General agreement that the test is warranted

The test is warranted with recurrent syncope or a single episode in a high-risk patient (e.g., pilot, surgeon, machine operator), whether or not the medical history is suggestive of neurally mediated origin, and with no evidence of SHD or if SHD is present but other causes of syncope have been excluded by appropriate testing.

Further evaluation is warranted of patients in whom an apparent cause has been established (e.g., asystole, AV block) but in whom demonstration of susceptibility to neurally mediated syncope would affect treatment plans.

Finally, TTT is indicated as part of the evaluation of exercise-induced or exercise-associated syncope.

2. Conditions in which differences of opinion regarding the utility of TTT remain

These include the following: differentiating convulsive syncope from seizures; evaluating patients (especially the elderly) with recurrent, unexplained falls; assessing recurrent dizziness or presyncope; evaluating unexplained syncope in the setting of peripheral neuropathies or dysautonomias; and follow-up evaluation to assess therapy for neurally mediated syncope (that is, it may be helpful in assessing the extent to which a particular therapy is effective in preventing syncope, including temporary dual-chamber pacing prior to permanent pacemaker implantation).

3. Indications of potential interest that require further study

These include recurrent idiopathic vertigo suggestive of neurally mediated hypotension and/or bradycardia, recurrent transient ischemic attacks, chronic fatigue syndrome, and survivors of sudden infant death syndrome.

TTT is generally not indicated for a single syncopal episode without injury that occurs in a non- high-risk setting with clear-cut vasovagal clinical features, or when an alternative specific cause has been established and the additional demonstration of a neurally mediated susceptibility would not alter treatment plans.

TTT should be avoided in the presence of severe LV outflow tract obstruction, critical mitral stenosis, or severe coronary or cerebrovascular disease.

B. Technique

TTT technique includes:

- **Laboratory:** Quiet, dim lighting, comfortable temperature, 20 to 45 minutes of supine equilibration period.
- **Patient:** Fasting for several hours before the procedure, IV fluid replacement.
- **Recordings:** At least 3 ECG leads, frequent BP testing.
- **Table:** Foot board support, smooth and rapid transitions.
- **Tilt:** Angle 70° to 80°, with initial drug-free tilt duration of 30 to 45 minutes.
- **Pharmacological provocation:** Isoproterenol (10 mcg/kg/min), nitroglycerine (400 mcg SL), or adenosine triphosphate (20 mg IV)
- **Supervision:** Nurse or laboratory technician experienced in TTT and cardiovascular laboratory procedures; physician in attendance or in proximity and immediately available.

The test should be terminated if syncope occurs; if symptoms are reproduced and accompanied by a substantial drop in BP, HR, or double product (e.g., ≤ 9000); or if the patient experiences chest pain, nausea, or dyspnea.

If repeated studies are required, they should ideally be performed under similar conditions, including time of day.

C. Positive Responses to TTT

The type of abnormal response to TTT depends on the predominance of cardioinhibitory or vasodepressor components; the modified VASIS classification is widely used:

- **Type 1 (mixed):** HR falls at the time of syncope, but the ventricular rate does not fall < 40 bpm, or falls < 40 bpm for < 10 seconds with or without asystole of < 3 seconds. BP falls before the HR falls.
- **Type 2A (cardioinhibition without asystole):** HR falls to a ventricular rate < 40 bpm for > 10 seconds, but asystole of > 3 seconds does not occur. BP falls before the HR falls.
- **Type 2B (cardioinhibition with asystole):** Asystole occurs for > 3 seconds. HR fall coincides with or precedes BP fall.
- **Type 3 (vasodepressor):** HR does not fall > 10% from maximal rate during tilt.

- **Exception 1 (chronotropic incompetence):** No increase in HR during tilt, or an increase < 10% from the pretilt rate.
- **Exception 2 (excessive HR increase):** Excessive HR rise both at the onset of the upright position and through its duration before syncope (> 130 bpm).

Patients who may benefit from cardiac pacing include those with type 2 responses, particularly 2B, and exception 1.

D. Complications

The induction of hypotension/bradycardia, with presyncope/syncope, is the intended end result of the test. If they occur, the patient should be returned to the supine position and attended to immediately (occasionally IV fluids, atropine, or adrenaline is required). In very rare cases, seizures or prolonged asystole may result; patients should be managed accordingly.

VII. AUTONOMIC MODULATION TESTING (OTHER THAN TTT)

There is an intricate and complex relationship between cardiovascular and autonomic nervous systems. For example, increased sympathetic and decreased parasympathetic activity enhances susceptibility to ventricular arrhythmias and sudden death, and β-blockers provide protective effects. Many markers of autonomic activity have been described, and most seem to correlate with an adverse prognosis in patients with SHD. While numerous tests are available, their utility alone or in combination remains to be clearly defined.

A. Heart Rate Variability

Heart rate variability (HRV) refers to the oscillation in the time interval between consecutive heart beats and reflects overall autonomic modulation. Reduced HRV has been associated with adverse prognosis in several disease states including post-MI, cardiomyopathy, HF, diabetes, amyloidosis, and sleep apnea. Importantly, a casual relationship between low HRV and mortality has not been established.

HRV can be measured by several methods:

1. Time domain analyses

Time domain measures are based on statistical calculations of R-R intervals. They are not interchangeable, as they do not necessarily reflect similar physiology. The most commonly utilized methods are standard deviation of normal-to-normal R-R intervals over a 24-hour period (SDNN) or over a brief (usually 5-min) period; standard deviation of the average normal-to-normal intervals for 288 5-minute intervals recorded by continuous 24-hour ECG monitoring (SDANN); percentage of normal R-R intervals that differ by > 50 milliseconds (pNN50); and the root mean square of successive R-R intervals (rMSSD).

2. Frequency domain analyses

Frequency domain measures use spectral analysis of a sequence of R-R intervals to provide information on how power (variance) is distributed as a function of frequency. Recordings are usually very short (3–5 min).

Commonly used methods to analyze frequency domain are total power, ultra-low-frequency (ULF) power, very-low-frequency (VLF) power, low-frequency (LF) power, high-frequency (HF) power, and the ratio of LF:HF. Changes in the LF band spectral power (0.04- to 0.15-Hz frequency range) reflect a combination of sympathetic and parasympathetic autonomic nervous system outflow variations, while changes in the HF band spectral power (0.15- to 0.40-Hz range) reflect vagal modulation of cardiac activity. The LF:HF power ratio is used as an index for assessing sympatho-vagal balance.

3. Nonlinear/complexity-based measures

These newer techniques study the structure of variability independent of the scale (e.g., chaotic analysis, approximate entropy, detrended fluctuation analysis).

No single method has been clearly shown to be superior. Rather, they appear complementary. Free software for HRV analysis and open access databases are available at www.physionet.org.

B. Baroreflex Sensitivity

Baroreflex sensitivity (BRS) refers to the vagally mediated reflex between baroreceptors and the HR, clinically expressed as the adaptation of R-R intervals to changes in BP. A reduced BRS reflects a reduction in parasympathetic activity, with a concomitant increase in sympathetic activity. It has been reported to have independent prognostic value for cardiac mortality following MI and in patients with HF, with increased risk for ventricular tachyarrhythmias.

BRS may be evaluated using several methods: vasoactive (α-agonists) drugs; the Valsalva maneuver; the neck chamber technique; the sequence method, which analyzes the relationship between increasing/decreasing ramps of BP and related increasing/decreasing changes in R-R intervals through linear regression; and spectral methods, which assess the relationship between specific oscillatory components of the signals. Once the reflex provocation is obtained, careful analysis of simultaneous recordings of ECG R-R intervals and systolic BP values allow characterization of BRS, expressed as the slope of the regression line relating dependency of R-R intervals to BP. Specific responses depend on the provocation method employed (e.g., the normal response to 25–100 μg of IV phenylephrine is a > 20 mm Hg increase in systolic BP and R-R interval prolongation >10 milliseconds for each 1 mm Hg increase in BP).

Despite the supporting data, further studies with larger patient populations are needed to firmly establish the clinical utility of BRS.

C. Heart Rate Turbulence

Heart rate turbulence (HRT) describes the response of the sinus node to premature ventricular contractions (PVCs) and consists of initial acceleration followed by deceleration. It is postulated that HRT is a measure of vagal responsiveness akin to BRS. It has been associated with SCD following MI (with or without ventricular dysfunction) and in patients with a history of VT or VF.

Two measures are used to quantify changes in the R-R interval post-PVC: turbulence onset (TO), which refers to the amount of sinus acceleration following a PVC, and turbulence slope (TS), which is the rate of sinus deceleration that follows acceleration. Strong sinus acceleration followed by rapid deceleration is the hallmark of a normal response, with TO < 0.0 and TS > 2.5.

Using both criteria, HRT us associated with a sensitivity of 30% and a positive predictive accuracy of approximately 30% as well. It, therefore, appears to be of limited value as a stand-alone test but may be helpful for risk stratification when combined with other predictors.

Other HRT indices have been proposed (e.g., turbulence dynamicity, turbulence frequency decrease, turbulence timing, turbulence jump, and correlation coefficient of TS), but they all require further study.

HRT is usually obtained from averaged R-R intervals pre- and post-PVCs over a 24-hour Holter recording, or from analysis of event records of ICDs or permanent pacemakers. It can also be induced by EPS or in patients with ICDs/pacemakers (*induced HRT*), using a conventional protocol (e.g., average of R-R intervals following 10 automated extra-stimuli with at least 20 intervening sinus beats and a coupling interval that is 60%–70% of the sinus CL).

Information on HRT may also be gathered from a 12-lead ECG. It has been proposed that HRT is abnormal if the R-R interval after a random PVC increases by at least 12 milliseconds from the shortest to the longest R-R of 10 consecutive beats.

VIII. SIGNAL-AVERAGED ECG

A. General Principles

The SAECG is a noninvasive ECG technique that averages multiple electrical signals from the heart to remove interference so as to detect small variations in the QRS complex. It may characterize abnormalities in the form of low-amplitude potentials in the terminal portion. These "late potentials" reflect areas of slow conduction that may be substrates for ventricular tachyarrhythmias.

The "noise" in a regular ECG ranges from 8 to 10 μV, generated primarily by skeletal muscle activity. Signal averaging improves the signal to noise ratio to facilitate the detection of the late potentials, with modern systems reducing noise to < 1 μV. Signals can be averaged by temporal and spatial techniques. Most current systems use time-domain analysis to detect late potentials, usually with bidirectional filtering and signal processing. The resulting filtered QRS complex is then analyzed in terms of duration, root-mean-square voltage of the terminal 40 milliseconds, and the duration that the filtered QRS complex remains < 40 μV.

Characteristics of a late potential, using a 40-Hz high-pass bidirectional filter, are filtered QRS complex > 114 milliseconds, a signal > 20 μV in the last 40 milliseconds of the filtered QRS, and voltage < 40 μV in the terminal QRS complex for > 38 milliseconds.

By using a Fourier transform, a sequence generated by sampling a time-domain signal can also be represented in the frequency domain; this technique is not widely available and end points of spectral analysis are still lacking.

B. Indications and Technique

The SAECG testing is carried out using regular ECG recording techniques with built-in equipment and software. The test usually takes between 5 and 20 minutes and reports are readily available.

High negative and low positive predictive values have been reported, although these measures are highly dependent on the prevalence of disease within the screened patient population. Though it may identify patients with prior MI at risk for SCD (mostly in those at low risk given its high negative predictive value), in general SAECG appears most helpful when considered as one of several diagnostic criteria or risk factors. It is not considered yet a standard technique.

IX. T-WAVE ALTERNANS TESTING

A. General Principles

T-wave alternans (TWA) testing is a noninvasive technique that considers beat-to-beat variations in the amplitude and morphology of T waves. It has also been associated with SCD due to ventricular tachyarrhythmias. Alternans of T waves result from repolarization abnormalities, most likely arising when the HR exceeds the capacity of cardiac cells to cycle intracellular Ca^{2+}. As a rate-dependent phenomenon, it tends to occur at lower HR in patients susceptible to ventricular arrhythmias. Some investigators have suggested that its predictive value rivals or exceeds EPS and LV EF. Its negative predictive value has been reported to be as high as 98%.

Indications for TWA testing remain to be defined, as it has not yet become standard of care. Proponents of TWA have suggested that it may be of value for the following clinical scenarios: presence of unexplained syncope/presyncope with known or suspected SHD, or abnormal ECG, or occurring suddenly or with exertion, or with risk factors for CAD; history of syncope/presyncope or complex ectopy and suspicion of CAD or in the setting of a family history of sudden death; sustained VT or VF associated with a transient or reversible cause such as ischemia, cardiac surgery, drug overdose, and so forth; suspected or documented NSVT and LV dysfunction; and symptomatic cardiac arrhythmias of undetermined origin.

B. Technique

The test quantifies beat-to-beat variability in the vector and amplitude of T waves. Detection of TWA requires graded exercise, special electrodes, and digital signal processing techniques to accurately record microvolt changes. Two methods currently exist: the spectral analytic technique (most common) and the modified moving average method. For testing to be accurate, the patient should be in sinus tachycardia with a ventricular rate > 105 bpm. It is of lesser value in patients with AF or in those who cannot sustain a HR of at least 105 bpm. In the post-MI setting, TWA should not be performed within the first 3 to 4 weeks, since fluctuations vary with the healing process.

TWA results are categorized as positive (> 1.9 μV of alternans beginning at a HR < 110 bpm), negative, or indeterminate. Patients with indeterminate results may undergo repeat testing. However, positive and indeterminate results are often grouped together, with recommendations to pursue further investigations that often lead to therapy (e.g., AAD, ICD implant).

As with other emerging markers of risk, further studies are needed regarding the clinical utility of TWA testing and its role in risk stratification schemes.

X. ANTICOAGULATION IN EP

Anticoagulation is extensively used in cardiac electrophysiology, either as long-term therapy to prevent/treat thromboembolic complications (described elsewhere in this book) or preprocedurally, as discussed next.

A. Ablation Procedures

The management of preprocedural anticoagulation is variable. In general, patients already on oral anticoagulants should discontinue therapy for at least 48 hours prior to the intervention. In patients at high risk of stroke (e.g., mechanical heart valve, prior history of stroke), bridging with LMWH or standard unfractionated heparin is recommended. While LMWH has the advantage of ease of administration and

Table 13.2. Considerations before Deciding Use of LMWH

Type and timeline of the procedure
Thromboembolic risk
Hemorrhagic risk
Weight (obesity)
Renal function and baseline platelet count
Instructions on performing self-injections
Detailed and patient-specific bridging schedules that outline:

- When to stop oral anticoagulation
- When to start and stop LMWH
  - 12 h prior to percutaneous procedures
  - 24 h prior to device implantation
- Timing of postprocedure INRs

out-of-hospital use, it has not been adequately evaluated in this setting. Several factors, summarized in Table 13.2, should be taken into consideration.

Bridging with heparin is generally not required in patients with lone AF or with bioprosthetic valves. Aspirin is not usually discontinued. Depending on the clinical situation, discontinuation of clopidrogel should be considered, ideally 7 days prior to the intervention.

Hemorrhagic and thromboembolic risks are mediated, in part, by the type of ablation procedure.

1. Low-risk ablation procedures

With right-sided interventions (e.g., ablation of AT, right-sided accessory pathways, AVNRT, VT, and paroxysmal isthmus-dependent AFL with ablation in sinus rhythm), the thromboembolic risk is generally low and of little consequence. Bleeding risks with anticoagulation may outweigh thromboemblic risks. While some centers routinely administer a single bolus of IV heparin after having obtained venous access, others refrain from anticoagulation altogether.

2. Intermediate-risk ablation procedures

With, for example, persistent isthmus-dependent AFL, the thromboembolic risk (e.g., atrial thrombi, echocardiographic contrast, electrical stunning postrestoration of sinus rhythm) appears to be higher than standard right-sided ablation procedures but lower than AF. An anticoagulation regimen is proposed in Table 13.3.

With left-sided interventions excluding AF and patients with overt SHD (e.g., left-sided accessory pathway), anticoagulation preprocedure is not advised. IV heparin is initiated immediately after arterial or transseptal puncture. Postprocedure, anticoagulation strategies differ with little data to guide practice. Aspirin 80 to 325 mg/day may be prescribed for 4 to 8 weeks following ablation of accessory

Table 13.3. Anticoagulation for Ablation of Persistent Typical AFL

- Effective oral anticoagulation for 3–4 weeks or transesophageal echocardiography prior to the ablation procedure
- Stop warfarin 3–5 days before the procedure, bridging with heparin until 12 h (LMWH) or 4 h (standard heparin) before the ablation
- IV heparin during the procedure (evidence for this practice is lacking)
- Restart effective oral anticoagulation (8–12 h postablation) for 4–6 weeks postprocedure

pathways or focal tachyarrhythmias. In the event of extensive lines of ablation, IV heparin is generally administered for 12 to 24 hours and initiated 4 to 6 hours after the intervention. Aspirin and/or warfarin may be prescribed for 1 to 3 months.

3. High-risk ablation procedures

With, for example, AF, ablation carries a thromboembolic risk as high as 5%. A preprocedural transesophageal echocardiography is recommended to ensure the absence of intracardiac thrombus. Clopidrogrel should be discontinued, if possible, to reduce the risk of tamponade in the event of cardiac perforation.

An anticoagulation scheme is proposed in Table 13.4.

As another example, catheter ablation of left-sided VT in the presence of SHD is associated with a 2.8% rate of thromboembolic complications. Echocardiographic screening is recommended to rule out the presence of intracardiac thrombus. A suggested anticoagulation protocol is outlined in Table 13.5.

B. Implantable Devices

Anticoagulation practices in the setting of implantable devices are evolving in response to new data regarding risks of pocket hematomas and thromboembolic

Table 13.4. Anticoagulation for Ablation of AF

- Effective oral anticoagulation for at least 4 weeks preprocedure
- Stop warfarin 4–5 days before the procedure, bridging with heparin until 12 h (LMWH) or 4 h (standard heparin) before the ablation
- IV heparin after sheath insertion and transseptal puncture
- Standard IV heparin restarted for 12–24 h postprocedure, with effective oral anticoagulation resuming the day of the procedure
- Replace IV heparin with LMWH after 12–24 h
- Stop LMWH when target INR (2–3) is reached
- Continue therapeutic warfarin for at least 3 months postprocedure

Table 13.5. Anticoagulation for Ablation of VT with SHD

- If prior oral anticoagulation: stop warfarin 4–5 days before the procedure, bridging with heparin until 12 h (LMWH) or 4 h (standard heparin) before the ablation
- IV heparin after sheath insertion and until the end of the ablation
- Standard IV heparin restarted (with ACT subtherapeutic) for 12–24 h postprocedure, with effective oral anticoagulation resuming the day of the procedure if previously indicated
- Replace IV heparin with LMWH after 12–24 h
- Stop LMWH when target INR (2–3) is reached
- Continue therapeutic warfarin as per original indication or for at least 1 month postprocedure
- If no prior indication for oral anticoagulation: substitute warfarin for aspirin 80–325 mg/day for at least 1 month

events. The general tendency is toward avoiding IV heparin postintervention to the greatest degree possible, since it is strongly associated with hemorrhagic complications. In patients at low risk of thromboembolic events, oral anticoagulants should be discontinued, heparin bridging avoided, and oral anticoagulation gradually reinitiated following the intervention. In patients at higher risk of thromboembolic events, cardiologists are increasingly performing interventions under full anticoagulation with oral agents so as to avoid postprocedural heparin bridging. LMWH should be entirely avoided due to high bleeding risk postsurgery. In patients at high risk of thromboembolic complications and subtherapeutic INR levels, unfractionated heparin may be used postprocedurally to target a partial thromboplastin time between 60 and 80 seconds, or at least twice the baseline value. Heparin boluses should be avoided whenever possible. Warfarin may be administered along with unfractionated heparin until the INR becomes therapeutic. Compressive dressings for 24 to 48 hours may be helpful in reducing the incidence of hematomas.

SUGGESTED READINGS

Blanc JJ, Almendral J, Brignole M, et al. Consensus document on antithrombotic therapy in the setting of electrophysiological procedures. *Europace.* 2008;10(5): 513-527.

Brignole M, Menozzi C, Del Rosso A, et al. New classification of haemodynamics of vasovagal syncope: beyond the VASIS classification. Analysis of the pre-syncopal phase of the tilt test without and with nitroglycerin challenge. *Europace.* 2000;2:66-69.

Cain ME, Anderson JL, Arnsdof MF, Mason JW, Scheiman MM, Waldo AL. ACC expert consensus document: signal-averaged electrocardiography. *J Am Coll Cardiol.* 1996;27(1):238-249.

Goldberger JJ, Cain ME, Hohnloser SH, et al. AHA/ACCF/HRS scientific statement on noninvasive risk stratification techniques for identifying patients at risk for sudden cardiac death. *J Am Coll Cardiol.* 2008;52:1179-1199.

Piazza G, Goldhaber SZ. Periprocedural management of the chronically anticoagulated patient: critical pathways for bridging therapy. *Crit Pathways Cardiol.* 2003;2:96-103.

Watanabe MA. Heart rate turbulence: a review. *Indian Pacing and Electrophysiol J.* 2003;3(1):10-22.

Index

A figure is denoted by *f*; a table by *t*.